Springer Theses

Recognizing Outstanding Ph.D. Research

Aims and Scope

The series "Springer Theses" brings together a selection of the very best Ph.D. theses from around the world and across the physical sciences. Each volume has been selected for its scientific excellence by an internationally renowned institute or department. For greater accessibility to non-specialists, the published versions include an extended introduction, as well as a foreword by the student's supervisor explaining the special relevance of the work for the field. The series as a whole provides a valuable resource both for newcomers to the research fields, and for other scientists seeking detailed background information on special questions. Finally, it serves as an accredited documentation of the valuable contributions made by today's younger generation of scientists.

How to Apply

Theses prepared at internationally leading universities and institutes may be nominated for publication by supervisors with the endorsement of the head of department. Each such department is entitled to nominate one thesis per year. Enquiries and requests for a nomination form should be sent, together with the thesis, to Ms. Nathalie Berg (see link on right).

Eligibility Criteria

To qualify for consideration, theses should fulfil the following criteria:

- They must be written in good English.
- The topic should fall within the confines of Chemistry, Physics, Earth Sciences, Engineering and related interdisciplinary fields such as Materials, Nanoscience, Chemical Engineering, Complex Systems and Biophysics.
- The work reported in the thesis must represent a significant scientific advance.
- A maximum of 30% of the text may be a verbatim reproduction from the author's previous publications.
- They must have been examined and passed during the 12 months prior to nomination.
- The theses should have a clearly defined structure including an introduction accessible to new PhD students and scientists not expert in the relevant field.
- After acceptance, theses should be supplemented by a "Supervisor's Foreword" outlining the wider scientific relevance of the work

Indexed by zbMATH.

Xingzhuo Chen

Deep Learning on Type Ia Supernovae

Doctoral thesis accepted by Texas A&M University, USA

Springer

Xingzhuo Chen
Institute of Data Science
Texas A&M University
College Station, TX, USA

ISSN 2190-5053 ISSN 2190-5061 (electronic)
Springer Theses
ISBN 978-3-032-13031-0 ISBN 978-3-032-13032-7 (eBook)
https://doi.org/10.1007/978-3-032-13032-7

This Springer imprint is published by the registered company Springer Nature Switzerland AG
The registered company address is: Gewerbestrasse 11, 6330 Cham, Switzerland

To my family.

To my family

Supervisor's Foreword

The emerging artificial intelligence (AI) revolution has spread to the field of scientific research. In astronomy, it has evolved into tools that are essential in making breakthrough discoveries, from exoplanet detections in the Milky Way galaxies to cosmic transients out to the distant galaxies. The doctoral dissertation titled "Artificial Intelligence Assisted Inversion on Type Ia Supernovae" by Xingzhuo Chen, submitted to Texas A&M University in 2024, explores novel computational methods—primarily based on AI—to study Type Ia supernovae (SNe Ia), which are vital tools in astrophysics and cosmology due to their role as standardizable candles for measuring cosmic distances.

The objectives of the dissertation are to improve the modeling, analysis, and understanding of Type Ia supernovae by integrating AI methods with traditional radiative transfer and spectral analysis tools, thus enabling better estimation of supernova properties and explosion mechanisms. This dissertation showcases how machine learning, particularly deep neural networks and physics-informed models, can transform the analysis of supernovae.

The first part of the dissertation is on AI-assisted inversion (AIAI) of the radiative transfer code TARDIS. The AI-assisted methods introduced allow for faster and more accurate inversion of observed spectra to derive the kinematics and chemical compositions of the supernova ejecta that can best match the observations. This inverse modeling approach makes theoretical radiative transfer modeling transparent from the observations to the ejecta structure. This transparency leads to a deeper insight into the elemental synthesis and explosion mechanisms of supernovae.

The second part is based on the SEDONA code. It introduces an Integral Based Technique (IBT) to accelerate Monte Carlo simulations of radiative transfer, improving computational efficiency. The method is validated against existing methods for accuracy and speed, enabling faster modeling of SN spectra. This improvement is important for 3-D modeling of the radiative process and the observational tests of the ejecta geometry of supernovae with spectropolarimetric data.

In the third part, AI method is applied to observational data that improves the understanding of supernova delay times and progenitor demographics.

Artificial intelligence (AI)-assisted modeling represents a transformative approach to scientific discovery by integrating complex theoretical frameworks with large-scale observational datasets. In the context of astrophysics, and particularly in the study of Type Ia supernovae, this synergy enables more accurate and efficient analysis of phenomena that were previously constrained by computational limitations. By leveraging advancements in deep learning and data-driven modeling—much like the innovations seen in the tech industry's big-data revolution—AI provides powerful tools for extracting physical insights from high-dimensional data. This approach not only accelerates simulations and inversion processes but also opens new avenues for uncovering patterns and correlations that may remain hidden in traditional methodologies. As astronomical data volumes continue to grow, AI-driven science is poised to play a pivotal role in shaping the future of theoretical and observational research.

Texas A&M University, College Station, TX, USA Lifan Wang

Acknowledgments I would like to thank Prof. Lifan Wang. Thank you, Professor, for hosting me as a student in TAMU, and for your insightful edification in every projects we collaborated, everything will not happen without you.

I would like to thank my girlfriend Lanyin Luo, she taught me how to survive in this country and her love supports me to pursue the PhD degree.

I would like to thank all of my collaborators, including but not limited to Prof. Nicholas Suntzeff, Prof. Kevin Krisciunas, Prof. Peter Brown, Prof. Lucas Macri, Dr. Lisa Perez, Prof. Ulisses Braga-Neto, Prof. David Jeffery, Prof. Antonella Palmese, and Dr. Zhong Ming. Each of you have taught me different things, including physics, computer science, astronomical observation, and history, during my PhD period.

I would like to thank TAMU, HPRC, and NERSC, for the cutting-edge computation platforms and maintenance.

I would like to thank Dr. Lei Hu, Dr. Tianrui Sun, Dr. Maokai Hu, Dr. Shihao Kou, and Prof. Xuewen Liu. You enlightened me in this field.

I would like to thank all my classmates in TAMU for their friendship and help.

Lastly, I would like to thank my family for their support and encouragement to pursue the PhD degree.

Contents

Contributors and Funding Sources

Contributors

This work was supported by a dissertation committee consisting of Prof. Lifan Wang, Lucas Macri, and Peter Brown of the George P. and Cynthia Woods Mitchell Institute for Fundamental Physics & Astronomy, Texas A.&M. University, Department of Physics and Astronomy, and Dr. Lisa Perez of the High Performance Research Computing, Texas A.&M. University.

The HST ultraviolet observational spectra in Chapter 2 was provided by Dr. Kate Magurie. Other observational spectra in Chapter 2 was provided by various projects, available for download on WISeREP website.

The MUSE spectrograph data in Chapter 4 was provided by 37 VLT observation programs (099.B-0193(A), 097.B-0776(A), 095.D-0091(A), 1100.B-0651(A), 095.B-0532(A), 096.B-0230(A), 60.A-9304(A), 099.D-0022(A), 0101.B-0706(A), 196.B-0578(B), 095.B-0624(A), 096.B-0325(A), 60.A-9100(G), 098.D-0115(A), 0103.D-0440(A), 099.B-0294(A), 099.B-0137(A), 097.B-0640(A), 0101.D-0748(B), 096.D-0296(A), 196.B-0578(A), 0102.B-0048(A), 094.B-0298(A), 0101.D-0748(A), 0100.D-0341(A), 095.D-0091(B), 097.A-0366(A), 60.A-9339(A), 097.D-0408(A), 098.B-0551(A), 099.A-0870(A), 094.A-0205(B), 296.B-5054(A), 094.B-0733(B), 096.D-0263(A), 0103.A-0637(A), 0102.D-0095(A)).

The SDSS MaNGA spectrograph data is provided by Sloan Digital Sky Survey. Funding for the Sloan Digital Sky Survey V has been provided by the Alfred P. Sloan Foundation, the Heising-Simons Foundation, the National Science Foundation, and the Participating Institutions. SDSS acknowledges support and resources from the Center for High-Performance Computing at the University of Utah. SDSS telescopes are located at Apache Point Observatory, funded by the Astrophysical Research Consortium and operated by New Mexico State University, and at Las Campanas Observatory, operated by the Carnegie Institution for Science. The SDSS web site is www.sdss.org. SDSS is managed by the Astrophysical Research Consortium for the Participating Institutions of the SDSS Collaboration, including Caltech, The Carnegie Institution for Science, Chilean National Time Allocation Committee (CNTAC) ratified researchers, The Flatiron Institute, the Gotham Participation Group, Harvard University, Heidelberg University, The Johns Hopkins University, L'Ecole polytechnique fédérale de Lausanne (EPFL), Leibniz-Institut für Astrophysik Potsdam (AIP), Max-Planck-Institut für Astronomie (MPIA Heidelberg), Max-Planck-Institut für Extraterrestrische Physik (MPE), Nanjing University, National Astronomical Observatories of China (NAOC), New Mexico State University, The Ohio State University, Pennsylvania State University, Smithsonian Astrophysical Observatory, Space Telescope Science Institute (STScI), the Stellar Astrophysics Participation Group, Universidad Nacional Autónoma de México, University of Arizona, University of Colorado Boulder, University of Illinois at Urbana-Champaign, University of Toronto, University of Utah, University of Virginia, Yale University, and Yunnan University.

All other work conducted for the thesis (or) dissertation was completed by the student independently.

Funding Sources

The supernova research by Xingzhuo Chen, is supported by NSF grant AST-1817099.

Nomenclature

AI	Artificial Intelligence
DD	Double Degenerate
IFU	Integral Field Unit
MaNGA	Mapping Nearby Galaxies at Apogee Point Observatory
MCRT	Monte-Carlo Radiative Transfer
MUSE	Multiple Unit Spectroscopic Explorer
NLTE	Non-Local Thermodynamic Equilibrium
NN	Neural Network
PINN	Physics Informed Neural Network
SD	Single Degenerate
SDSS	Sloan Digital Sky Survey
SFH	Star Formation History
SN	Supernova
SN Ia	Type Ia Supernova
TAMU	Texas A&M University
VLT	Very Large Telescope

Chapter 1
Introduction and Literature Review

Abstract Type Ia supernovae (SNe Ia) play a crucial role in cosmology as standardizable candles due to empirical correlations between their light curve properties and intrinsic luminosities. These observations are broadly consistent with theoretical models involving the thermonuclear explosion of a carbon-oxygen white dwarf near the Chandrasekhar mass. However, first principles simulations incorporating hydrodynamics, nucleosynthesis, and radiative transfer remain computationally expensive and cannot yet fully reproduce observed features with high precision. This chapter reviews the explosion mechanisms of SNe Ia, focusing on radiative transfer modeling and progenitor system scenarios. The uncertainty in progenitor systems—whether single degenerate or double degenerate—motivates the use of delay time distribution (DTD) modeling and population synthesis to connect observed supernova (SN) rates with theoretical channels. Recent efforts leveraging host galaxy stellar populations and local environments have provided additional constraints on progenitor age and metallicity. Finally, we discuss the growing application of artificial intelligence (AI), particularly convolutional neural networks (CNNs) and physics-informed neural networks (PINNs), in stellar spectroscopy and radiative transfer problems. These AI approaches offer promising pathways to accelerate simulations and solve complex PDEs in a data-driven yet physically consistent manner. Together, these developments represent a multidisciplinary effort to improve our understanding of SNe Ia explosions and their progenitors.

1.1 Type Ia Supernova Explosion Mechanism

Type Ia supernovae (SNe Ia) are used for cosmological distance measurements based on empirical relations between their light curve shapes and luminosities [1–7]. These relations are consistent with theoretical models of the thermonuclear explosion of a white dwarf close to the Chandrasekhar mass limit (1.4 $M_{\odot}$). First principle hydrodynamics, nucleosynthesis, and radiative transfer simulations of SNe Ia are expensive and are still far from the precision that can match those of the observations, although some three-dimensional SNe Ia models can indeed provide

X. Chen, *Deep Learning on Type Ia Supernovae*, Springer Theses,
https://doi.org/10.1007/978-3-032-13032-7_1

photometric and spectroscopic fits that are comparable to those observed in SNe Ia (e.g., [8–10]).

1.2 Radiative Transfer

Hydrodynamic and nucleosynthesis processes are significant in the first $\sim$100 seconds after the explosion (e.g., [11]). Thereafter, the supernova (SN) ejecta expand homologously, and the observed spectra and light curves can be modeled accordingly. Modeling the radiative transfer process is essential in revealing the density profiles and elemental abundances of the SN ejecta to constrain the SN explosion mechanisms. Several codes have been developed to calculate synthetic SN spectra. One of the simplest is the SYNOW code [12, 13], which uses line opacities as input to generate synthetic spectra and has been widely used for spectral line identifications. Hydra [14], PHOENIX [15], and CMFGEN [16] are more advanced codes that allow for NLTE modeling through a finite element approach. Programs using the Monte Carlo method have also been developed for spectral simulation (e.g., SEDONA [17], and ARTIS [18]) and have been used for spectropolarimetry modeling [19–22]. Among these codes, TARDIS [23] is an one-dimensional, time-independent Monte Carlo radiative transfer (MCRT) program with approximate NLTE treatment of atomic processes. The SN ejecta structure is assumed to be spherical symmetric, and a lower boundary is set in the simulation, in order to avoid the time-consuming gamma-ray photon propagation calculations. The optical energy packets emitted at the lower boundary propagate through the SN ejecta plasma and undergo physical processes that change the energy, frequency, and direction of the optical packets, including (1) bound-bound transition, (2) bound-free transition, (3) free-free transition, (4) and electron Thomson scattering events. The escaped optical energy packets are summed to synthesize the observable optical spectra. These radiative codes model the physical processes forwardly, starting from a given set of initial and boundary conditions. Kerzendorf et al. [24] used a neural network (NN) to accelerate the calculation of the radiative transfer program and suggest that the method can be used for SN ejecta structure estimates based on a nested sampling algorithm. In spectropolarimetry models, 3-D effects, nonlocal thermodynamic equilibrium (NLTE), and time-dependent radiative transfer effects are all important. SEDONA [17] is capable of 3-D time-dependent radiative transfer calculation with the inclusion of NLTE physical processes and spectropolarimetry calculations. The gamma-ray emission from radioactive elements is the energy source in the simulation. The gamma-ray energy packets undergo bound-free transition, Compton scattering, and photoelectric scattering process and then converted into optical energy packets. Due to the extra physical processes and the complexity of 3-D ejecta structure, the SEDONA simulation requires much more computation resources comparing to TARDIS.

1.3 SNe Ia Progenitors

SNe Ia are produced by the explosion of white dwarfs (WDs) in binary systems (e.g., [25, 26]). However, the configuration of the binary system remains unknown, and there are two leading scenarios. In the single-degenerate (SD) scenario, the progenitor WD accretes matter from a nondegenerate companion star to reach a critical mass $\sim 1.37 M_{\odot}$ before SN explosion [27, 28]. In the double-degenerate (DD) scenario, the explosion is triggered by the merging of two WDs [29, 30].

The delay time distribution (DTD), which describes the SN Ia rate per unit stellar mass as a function of the time after a burst of star formation activity, has been extensively used to investigate the SD and DD scenarios and the observed SN event statistics (e.g., [31, 32]). In binary population synthesis (BPS) simulations of carbon-oxygen white dwarfs (CO WD) and helium white dwarfs (He WD) merger [33, 34], the delay time (denoted as τ) from the star formation activity to the first SN Ia is typically $10^{8.5} \sim 10^9$ years, and the SN rate follows a t^{-1} decay. Similarly, the BPS simulations of CO WD and CO WD also show a t^{-1} relation, but the delay time of the first SN Ia is around $10^8 \sim 10^{8.5}$ years [34, 35]. In contrast, simulations for the SD scenario show different delay times for different channels. [36] and [37] simulated the CO WD and main sequence star (CO WD+MS) channel and found most of the SNe Ia are produced in $10^8 \sim 10^9$ years after star formation. Liu et al. [38] simulated the CO WD and red giant star (CO WD + RG) channel and found the delay time to be $10^{8.6} \sim 10^{8.7}$ years but with a steeper delay time relation than that of the DD scenario. Apart from MS or RG serving as the companion star, [39] simulated CO WD + He star as SN Ia progenitor system, which shows a peak of the event rate around 10^8 years. In addition, [40] proposed a new channel with carbon-oxygen-neon (CONe) WD serving as SNe Ia progenitor. In both the CONe WD + He star channel [41] and CONe WD + MS channel [42], the delay times are $10^{7.5} \sim 10^9$ years and do not show the t^{-1} decaying rate. Although different BPS simulations of DD scenario show SN rates in agreement with observations (e.g., [34, 36, 43, 44]), the observed SN population can still originate from a combination of multiple channels [45].

A few major observational measurements of the DTDs are provided by the SN Ia rates in different redshift bins in galaxy clusters [46, 47] and from large untargeted supernova survey projects [48–52]. Alternatively, SN remnants can be used to measure the SN rate [53]. Although the measured results show a t^{-1} SN rate decay which is consistent with the DD scenario, the exact parameters of the relation are strongly dependent on the details of the cosmic star formation history (CSFH) [54]. Furthermore, due to the incompleteness of high-redshift SNe Ia discoveries in high-redshift SN surveys, the delay time τ is still not very strongly constrained.

In quest of an independent estimate of the SN Ia delay time, [55] and [56] used the host galaxy stellar population as a proxy to estimate the DTD assuming that SN progenitors share the same formation history with the other stars in the host galaxy. Based on such an assumption, [57] utilized the *Hubble* Space Telescope (HST) to directly observe the stars close to 9-type Iax SNe and constrained the delay time using their nearby stellar ages as proxies. Furthermore, [58] utilized integral field unit (IFU) facilities to acquire the host galaxy spectra of 17 SN 1991bg-like SNe at the sites of the SN explosions, calculated the stellar populations within ~1 kpc of the SNe, and concluded that SN 1991bg-like SNe originate from an older stellar population than normal SNe Ia. Notably, as core-collapse SNe originate from younger stellar populations, their local stellar population distributions show distinctively younger stellar populations than those for SNe Ia and set strong constraints on the ages of Type II, Ib, Ic, IIb, and IIn SN progenitors [59]. In addition to the correlation between the SN rates and the local stellar population, [60] proposed to use the distance between the SNe and H II regions as an SN progenitor age indicator, and [61] discussed the relation between DTD and galactic iron accumulation history. The SN host galaxy star-forming region [62], SN environmental metallicity [63], and galaxy velocity field [64] are also discussed for their potential influences on the SN Ia rate.

1.4 Artificial Intelligence

Deep learning techniques have been developed for stellar spectroscopy [65]. For stellar spectroscopy, most of the spectral features are narrow and well separated, and the required network depth is minimal. Fabbro et al. [65] applied a CNN with two convolution layers and two fully connected layers to the infrared spectra data of APOGEE [66]. Liu et al. [67] applied a CNN with eight convolution layers and two fully connected layers to mineral Raman spectrum classifications. Bu et al. [68] utilized five-layered CNN (combined with support vector machine, SVM) to detect hot subdwarf stars from LAMOST DR4 spectra.

Physics-informed neural networks (PINNs) have emerged recently as a powerful addition to traditional numerical partial differential equation (PDE) solvers [69, 70]. The PINN approach is based on constraining the output of a deep neural network to satisfy a physical model specified by a PDE. Using neural networks as universal function approximators to solve PDEs had been proposed already in the 1990s [71, 72]. PINN capabilities at solving PDEs have been enhanced in many different ways since then by utilizing the expressive powers of deep neural networks, which are made possible by the recent advances in GPU-computing and training algorithms [73], as well as computational advances in automatic differentiation methods [74]. A significant advantage of PINNs over traditional time-stepping PDE solvers is that PINNs are mesh-less and can solve in space and time simultaneously. Combined with the regression capability of deep neural networks, PINNs are also suitable for PDE-related inverse problems. In [75], PINNs were applied to solve several simple monochromatic and polychromatic radiative transfer problems.

References

1. Phillips MM (Aug. 1993) The absolute magnitudes of type IA supernovae. Astrophys J Lett 413:L105
2. Riess AG, Filippenko AV, Challis P, Clocchiatti A, Diercks A, Garnavich PM, Gilliland RL, Hogan CJ, Jha S, Kirshner RP, Leibundgut B, Phillips MM, Reiss D, Schmidt BP, Schommer RA, Smith RC, Spyromilio J, Stubbs C, Suntzeff NB, Tonry J (Sept. 1998) Observational evidence from supernovae for an accelerating universe and a cosmological constant. Astronomical J 116:1009–1038
3. Perlmutter S, Aldering G, Goldhaber G, Knop RA, Nugent P, Castro PG, Deustua S, Fabbro S, Goobar A, Groom DE, Hook IM, Kim AG, Kim MY, Lee JC, Nunes NJ, Pain R, Pennypacker CR, Quimby R, Lidman C, Ellis RS, Irwin M, McMahon RG, Ruiz-Lapuente P, Walton N, Schaefer B, Boyle BJ, Filippenko AV, Matheson T, Fruchter AS, Panagia N, Newberg HJM, Couch WJ, T.S.C. Project (June 1999) Measurements of and Λ from 42 high-redshift supernovae. Astrophys J 517:565–586
4. Wang L, Goldhaber G, Aldering G, Perlmutter S (Jun 2003) Multicolor light curves of type Ia supernovae on the color-magnitude diagram: A novel step toward more precise distance and extinction estimates. Astrophys J 590:944–970
5. Rubin D, Knop RA, Rykoff E, Aldering G, Amanullah R, Barbary K, Burns MS, Conley A, Connolly N, Deustua S, Fadeyev V, Fakhouri HK, Fruchter AS, Gibbons RA, Goldhaber G, Goobar A, Hsiao EY, Huang X, Kowalski M, Lidman C, Meyers J, Nordin J, Perlmutter S, Saunders C, Spadafora AL, Stanishev V, Suzuki N, Wang L (Jan 2013) Precision measurement of the most distant spectroscopically confirmed supernova Ia with the hubble space telescope. Astrophys J 763:35
6. Yang J, Wang L, Suntzeff N, Hu L, Aldoroty L, Brown PJ, Krisciunas K, Arcavi I, Burke J, Galbany L, Hiramatsu D, Hosseinzadeh G, Howell DA, McCully C, Pellegrino C, Valenti S (Oct. 2022) Using 1991T/1999aa-like type Ia supernovae as standardizable candles. Astrophys J 938:83
7. Aldoroty L, Wang L, Hoeflich P, Yang J, Suntzeff N, Aldering G, Antilogus P, Aragon C, Bailey S, Baltay C, Bongard S, Boone K, Buton C, Copin Y, Dixon S, Fouchez D, Gangler E, Gupta R, Hayden B, Karmen M, Kim AG, Kowalski M, Küsters D, Léget PF, Mondon F, Nordin J, Pain R, Pecontal E, Pereira R, Perlmutter S, Ponder KA, Rabinowitz D, Rigault M, Rubin D, Runge K, Saunders C, Smadja G, Suzuki N, Tao C, Thomas RC, Vincenzi M (Oct. 2022) Bump Morphology of the CMAGIC Diagram. arXiv e-prints, arXiv:2210.06708
8. Bulla M, Sim SA, Kromer M, Seitenzahl IR, Fink M, Ciaraldi-Schoolmann F, Röpke FK, Hillebrandt W, Pakmor R, Ruiter AJ, Taubenberger S (2016) Predicting polarization signatures for double-detonation and delayed-detonation models of Type Ia supernovae. Mon Not R Astron Soc 462:1039–1056
9. Wilk KD, Hillier DJ, Dessart L (2017) Ejecta mass diagnostics of type Ia supernovae. Mon Not R Astron Soc 474:3187–3211
10. Townsley DM, Miles BJ, Shen KJ, Kasen D (Jun 2019) Double detonations with thin, modestly enriched helium layers can make normal type Ia supernovae. Astrophys J 878:L38
11. Röpke FK (Mar. 2005) Following multi-dimensional type Ia supernova explosion models to homologous expansion. Astron Astrophys 432:969–983
12. Parrent J, Branch D, Jeffery D (Oct. 2010) SYNOW: a highly parameterized spectrum synthesis code for direct analysis of SN spectra
13. Thomas RC, Nugent PE, Meza JC (Feb. 2011) SYNAPPS: data-driven analysis for supernova spectroscopy. Publ Astron Soc Pac 123:237
14. Hoeflich P, Khokhlov A, Wheeler JC, Phillips MM, Suntzeff NB, Hamuy M (Dec. 1996) Maximum brightness and postmaximum decline of light curves of type IA supernovae: a comparison of theory and observations. Astrophys J Lett 472:L81
15. Hauschildt PH, Baron E (May 2006) A 3D radiative transfer framework. I. Non-local operator splitting and continuum scattering problems. Astron Astrophys 451:273–284

16. Hillier DJ, Miller DL (Mar. 1998) The treatment of non-LTE line blanketing in spherically expanding outflows. Astrophys J 496:407–427
17. Kasen D, Thomas RC, Nugent P (Nov. 2006) Time-dependent Monte Carlo radiative transfer calculations for three-dimensional supernova spectra, light curves, and polarization. Astrophys J 651:366–380
18. Kromer M, Sim SA (2009) Time-dependent three-dimensional spectrum synthesis for Type Ia supernovae. Mon Not R Astron Soc 398:1809–1826
19. Höflich P (June 1991) Asphericity effects in scatterring dominated photospheres. Astron Astrophys 246:481
20. Kasen D, Nugent P, Wang L, Howell DA, Wheeler JC, Höflich P, Baade D, Baron E, Hauschildt PH (Aug. 2003) Analysis of the flux and polarization spectra of the type Ia supernova SN 2001el: exploring the geometry of the high-velocity ejecta. Astrophys. J. 593:788–808
21. Bulla M, Sim SA, Kromer M (June 2015) Polarization spectral synthesis for type Ia supernova explosion models. Mon Not R Astron Soc 450:967–981
22. Livneh R, Katz B (Jan. 2022) Polarization signatures of the head-on collision model for Type Ia supernovae: how much asymmetry is too much? Mon Not R Astron Soc 511:2994–3004
23. Kerzendorf WE, Sim SA (2014) A spectral synthesis code for rapid modelling of supernovae. Mon Not R Astron Soc 440:387–404
24. Kerzendorf WE, Vogl C, Buchner J, Contardo G, Williamson M, van der Smagt P (Apr. 2021) Dalek: a deep learning emulator for TARDIS. Astrophys J Lett 910:L23
25. Maguire K (2016) Type Ia supernovae. Springer International Publishing, Cham, pp 1–24
26. Branch D, Wheeler JC (2017) Supernova explosions. Springer, Berlin/Heidelberg
27. Whelan J, Iben J, Icko (Dec. 1973) Binaries and supernovae of type I. Astrophys J 186:1007–1014
28. Nomoto K (Feb. 1982) Accreting white dwarf models for type I supernovae. I - Presupernova evolution and triggering mechanisms. Astrophys J 253:798–810
29. Webbink RF (Feb. 1984) Double white dwarfs as progenitors of R Coronae Borealis stars and type I supernovae. Astrophys J 277:355–360
30. Iben I, Tutukov AV (May 1984) The evolution of binaries of moderate primordial mass (<∼10 M0) and the formation of transient and explosive objects due to accumulation instabilities in such binaries In: Woosley SE (ed) American Institute of Physics Conference Series, vol. 115. American Institute of Physics Conference Series, pp 11–30
31. Maoz D, Mannucci F (Jan. 2012) Type-Ia supernova rates and the progenitor problem: a review. Publ Astron Soc Aust 29:447–465
32. Maoz D, Mannucci F, Nelemans G (Aug. 2014) Observational clues to the progenitors of type Ia supernovae. Ann Rev Astron Astrophys 52:107–170
33. Meng X, Han Z (Jan. 2015) A pair of CO + He white dwarfs as the progenitor of 2005E-like supernovae? Astron Astrophys 573:A57
34. Liu D, Wang B, Han Z (2018) The double-degenerate model for the progenitors of Type Ia supernovae. Mon Not R Astron Soc 473:5352–5361
35. Chen X, Jeffery CS, Zhang X, Han Z (Aug. 2012) Some constraints on the lower mass limit for double-degenerate progenitors of type Ia supernovae. Astrophys J Lett 755:L9
36. Claeys JSW, Pols OR, Izzard RG, Vink J, Verbunt FWM (Mar. 2014) Theoretical uncertainties of the Type Ia supernova rate. Astron Astrophys 563:A83
37. Wang B, Ma X, Liu DD, Liu ZW, Wu CY, Zhang JJ, Han Z (Apr. 2015) Super-Eddington wind scenario for the progenitors of type Ia supernovae: binary population synthesis calculations. Astron Astrophys 576:A86
38. Liu D, Wang B, Ge H, Chen X, Han Z (Feb. 2019) The progenitors of type-Ia supernovae in semidetached binaries with red giant donors. Astron Astrophys 622:A35
39. Wang B, Podsiadlowski P, Han Z (Dec. 2017) He-accreting carbon-oxygen white dwarfs and Type Ia supernovae. Mon Not R Astron Soc 472:1593–1599
40. Denissenkov PA, Herwig F, Truran JW, Paxton B (July 2013) The C-flame quenching by convective boundary mixing in super-AGB stars and the formation of hybrid C/O/Ne white dwarfs and SN progenitors. Astrophys J 772:37

41. Wang B, Meng X, Liu DD, Liu ZW, Han Z (Oct. 2014) The hybrid CONe WD + He star scenario for the progenitors of type Ia supernovae. Astrophys J Lett 794:L28
42. Meng X, Podsiadlowski P (July 2014) The birth rate of SNe Ia from hybrid CONe white dwarfs. Astrophys J Lett 789:L45
43. Toonen S, Nelemans G, Portegies Zwart S (Oct. 2012) Supernova Type Ia progenitors from merging double white dwarfs. Using a new population synthesis model. Astron Astrophys 546:A70
44. Ruiter AJ, Belczynski K, Fryer C (July 2009) Rates and delay times of Type Ia supernovae. Astrophys J 699:2026–2036
45. Nelemans G, Toonen S, Bours M (Jan. 2013) Theoretical delay time distributions. In: Di Stefano R, Orio M, Moe M (eds) Binary paths to Type Ia supernovae explosions, vol 281. IAU Symposium, pp 225–231
46. Friedmann M, Maoz D (2018) The rate of Type-Ia supernovae in galaxy clusters and the delay-time distribution out to redshift 1.75. Mon Not R Astron Soc 479:3563–3581
47. Freundlich J, Maoz D (Apr. 2021) The delay time distribution of Type-Ia supernovae in galaxy clusters: the impact of extended star-formation histories. Mon Not R Astron Soc 502:5882–5895
48. Madgwick DS, Hewett PC, Mortlock DJ, Wang L (Nov 2003) Spectroscopic detection of type Ia supernovae in the Sloan Digital Sky Survey. Astrophys J 599:L33–L36
49. Graur O, Maoz D (2013) Discovery of 90 Type Ia supernovae among 700 000 Sloan spectra: the Type Ia supernova rate versus galaxy mass and star formation rate at redshift 0.1. Mon Not R Astron Soc 430:1746–1763
50. Frohmaier C, Sullivan M, Nugent PE, Smith M, Dimitriadis G, Bloom JS, Cenko SB, Kasliwal MM, Kulkarni SR, Maguire K, Ofek EO, Poznanski D, Quimby RM (June 2019) The volumetric rate of normal type Ia supernovae in the local Universe discovered by the Palomar Transient Factory. Mon Not R Astron Soc 486:2308–2320
51. Perrett K, Sullivan M, Conley A, González-Gaitán S, Carlberg R, Fouchez D, Ripoche P, Neill JD, Astier P, Balam D, Balland C, Basa S, Guy J, Hardin D, Hook IM, Howell DA, Pain R, Palanque-Delabrouille N, Pritchet C, Regnault N, Rich J, Ruhlmann-Kleider V, Baumont S, Lidman C, Perlmutter S, Walker ES (Jul 2012) Evolution in the volumetric type Ia supernova rate from the supernova legacy survey. Astron J 144:59
52. Rodney SA, Riess AG, Strolger L-G, Dahlen T, Graur O, Casertano S, Dickinson ME, Ferguson HC, Garnavich P, Hayden B, Jha SW, Jones DO, Kirshner RP, Koekemoer AM, McCully C, Mobasher B, Patel B, Weiner BJ, Cenko SB, Clubb KI, Cooper M, Filippenko AV, Frederiksen TF, Hjorth J, Leibundgut B, Matheson T, Nayyeri H, Penner K, Trump J, Silverman JM, U V, Bostroem KA, Challis P, Rajan A, Wolff S, Faber SM, Grogin NA, Kocevski D (Jun 2014) Type Ia supernova rate measurements to redshift 2.5 from candels: searching for prompt explosions in the early universe. Astron J 148:13
53. Maoz D, Badenes C (2010) The supernova rate and delay time distribution in the Magellanic Clouds. Mon Not R Astron Soc 407:1314–1327
54. Gal-Yam A, Maoz D (2004) The redshift distribution of type Ia supernovae: constraints on progenitors and cosmic star formation history. Mon Not R Astron Soc 347:942–950
55. Maoz D, Mannucci F, Brandt TD (2012) The delay-time distribution of Type Ia supernovae from Sloan II. Mon Not R Astron Soc 426:3282–3294
56. Heringer E, Pritchet C, van Kerkwijk MH (Sep 2019) The delay times of type Ia supernova. Astrophys J 882:52
57. Takaro T, Foley RJ, McCully C, Fong W-f, Jha SW, Narayan G, Rest A, Stritzinger M, McKinnon K (2020) Constraining Type Iax supernova progenitor systems with stellar population age dating. Mon Not R Astron Soc 493:986–1002
58. Panther FH, Seitenzahl IR, Ruiter AJ, Crocker RM, Lidman C, Wang EX, Tucker BE, Groves B (2019) Sn1991bg-like supernovae are associated with old stellar populations. Publ Astron Soc Aust 36:e031
59. Kuncarayakti H, Anderson JP, Galbany L, Maeda K, Hamuy M, Aldering G, Arimoto N, Doi M, Morokuma T, Usuda T (May 2018) Constraints on core-collapse supernova progenitors from explosion site integral field spectroscopy. Astron Astrophys 613:A35

60. Galbany L (Mar. 2017) Connecting supernovae with their environments. In: Arribas S, Alonso-Herrero A, Figueras F, Hernández-Monteagudo C, Sánchez-Lavega A, Pérez-Hoyos S (eds) Highlights on Spanish Astrophysics IX, pp 153–158
61. Maoz D, Graur O (Oct. 2017) Star formation, supernovae, iron, and α: consistent cosmic and galactic histories. Astrophys J 848:25
62. Galbany L, Stanishev V, Mourão AM, Rodrigues M, Flores H, García-Benito R, Mast D, Mendoza MA, Sánchez SF, Badenes C, Barrera-Ballesteros J, Bland-Hawthorn J, Falcón-Barroso J, García-Lorenzo B, Gomes JM, González Delgado RM, Kehrig C, Lyubenova M, López-Sánchez AR, de Lorenzo-Cáceres A, Marino RA, Meidt S, Mollá M, Papaderos P, Pérez-Torres MA, Rosales-Ortega FF, van de Ven G (Dec. 2014) Nearby supernova host galaxies from the CALIFA Survey. I. Sample, data analysis, and correlation to star-forming regions. Astron Astrophys 572:A38
63. Galbany L, Stanishev V, Mourão AM, Rodrigues M, Flores H, Walcher CJ, Sánchez SF, García-Benito R, Mast D, Badenes C, González Delgado RM, Kehrig C, Lyubenova M, Marino RA, Mollá M, Meidt S, Pérez E, van de Ven G, Vílchez JM (June 2016) Nearby supernova host galaxies from the CALIFA survey. II. Supernova environmental metallicity. Astron Astrophys 591:A48
64. Zhou L, Liang Y-C, Ge J-Q, Shao X, Chen X-Y, Deng L-C (Aug. 2019) The local properties of supernova explosions and their host galaxies. Res Astron Astrophys 19:121
65. Fabbro S, Bialek S, O'Briain T, Kielty CL, Jahandar F, Monty S, Venn KA (2017) An application of deep learning in the analysis of stellar spectra. Mon Not R Astron Soc 475:2978–2993
66. Nidever DL, Holtzman JA, Prieto CA, Beland S, Bender C, Bizyaev D, Burton A, Desphande R, Fleming SW, Pérez AEG, Hearty FR, Majewski SR, Mészáros S, Muna D, Nguyen D, Schiavon RP, Shetrone M, Skrutskie MF, Sobeck JS, Wilson JC (Nov 2015) The data reduction pipeline for the apache point observatory galactic evolution experiment. Astron J 150:173
67. Liu J, Gibson SJ, Mills J, Osadchy M (2019) Dynamic spectrum matching with one-shot learning. Chemom Intell Lab Syst 184:175–181
68. Bu Y, Zeng J, Lei Z (May 2018) Searching for hot subdwarf stars from the LAMOST spectra. III. Classifying the hot subdwarf stars from LAMOST DR4 using deep learning method. arXiv e-prints
69. Raissi M, Perdikaris P, Karniadakis G (2019) Physics-informed neural networks: a deep learning framework for solving forward and inverse problems involving nonlinear partial differential equations. J Comput Phys 378:686–707
70. Karniadakis GE, Kevrekidis IG, Lu L, Perdikaris P, Wang S, Yang L (2021) Physics-informed machine learning. Nat Rev Phys 3(6):422–440
71. Dissanayake MW, G M, Phan-Thien N (1994) Neural-network-based approximations for solving partial differential equations. Commun Num Methods Eng 10(3):195–201
72. Lagaris IE, Likas A, Fotiadis DI (1998) Artificial neural networks for solving ordinary and partial differential equations. IEEE Trans Neural Netw 9(5):987–1000
73. Abadi M, Barham P, Chen J, Chen Z, Davis A, Dean J, Devin M, Ghemawat S, Irving G, Isard M, Kudlur M, Levenberg J, Monga R, Moore S, Murray DG, Steiner B, Tucker P, Vasudevan V, Warden P, Wicke M, Yu Y, Zheng X (Nov. 2016) TensorFlow: a system for large-scale machine learning, in 12th USENIX symposium on operating systems design and implementation (OSDI 16) (Savannah, GA). USENIX Association, Berkeley, pp 265–283
74. Baydin AG, Pearlmutter BA, Radul AA, Siskind JM (Jan 2017) Automatic differentiation in machine learning: a survey. J Mach Learn Res 18:5595–5637
75. Mishra S, Molinaro R (Aug 2021) Physics informed neural networks for simulating radiative transfer. J Quant Spectrosc Radiat Transf 270:107705

Chapter 2
Artificial Intelligence Assisted Inversion on Type Ia Supernovae

Abstract We generate ∼100,000 model spectra of Type Ia Supernovae (SN Ia) to form a spectral library for the purpose to build an Artificial Intelligence Assisted Inversion (AIAI) algorithm for theoretical models. As a first attempt, we restrict our studies to time around B-band maximum and compute theoretical spectra with a broad spectral wavelength coverage from 2000 to 10,000 Å using TARDIS. Based on the library of theoretically calculated spectra, we construct the AIAI algorithm with a multi-residual convolutional neural network (MRNN) which backs out the contribution of different ionic species to the heavily blended spectral profiles of the theoretical spectra. The AIAI is found to be very powerful in distinguishing spectral patterns due to coupled atomic transitions and has the capacity of measuring the contributions from different ionic species. By applying the AIAI algorithm to a set of well-observed SN Ia spectra, we demonstrate that the model can yield powerful constraints on the chemical structures of these SN Ia. Using the chemical structures deduced from AIAI, we successfully reconstructed the observed data, thus confirming the validity of the method. We show that the light curve decline rate of SN Ia is correlated with the amount of ^{56}Ni above the photosphere in the ejecta. We detect a clear decrease of ^{56}Ni mass with time that can be attributed to its radioactive decay.

2.1 TARDIS MCRT Calculation

Our study needs a library of SN spectra. This library spectra should capture the radiative processes involved in SNe Ia as much as possible and cover a broad range of the physical properties of the ejecta. The parameter space is large, and the number of models to be calculated can be huge. Codes that are very CPU demanding are apparently inappropriate for this study, at least for the current attempt which is still

---

Reprinted with permission from "Artificial Intelligence Assisted Inversion (AIAI) of Synthetic Type Ia Supernova Spectra" by Xingzhuo Chen, Lei Hu, Lifan Wang. 2020. The Astrophysical Journal Supplement Series, Vol 250, Issue 1, Number 12. https://doi.org/10.3847/1538-4365/ab9a3b. Copyright ©2020 The American Astronomical Society, reproduced with permission.

X. Chen, *Deep Learning on Type Ia Supernovae*, Springer Theses,
https://doi.org/10.1007/978-3-032-13032-7_2

an exploratory first step. For the determination of chemical structures, the primary requirement is that the code should be approximately correct in producing physical models to the spectral profiles of the most important observable ionic species. Luckily, there are several codes that fit these criteria. For example, the code SYNOW [1, 2] can run at very high speed and generate spectral libraries of a broad range of ejecta parameters. However, the code only allows a very crude description of the ejecta geometry. The input parameters are given in terms of optical depth of some reference lines of certain ionic species. Quantitative constraints to parameters related to ejecta structure and supernova luminosity is difficult using the available versions of SYNOW.

With the computing power available to us, the Monte-Carlo code TARDIS is a good compromise that matches our interest. The input parameters for TARDIS include the elemental abundances and density structures of the ejecta which can be flexibly modified. A spectrum can be calculated for any given day during the photospheric phase, once the location of the photosphere and the luminosity at that day are provided. TARDIS first calculates the electron density, level population of ions and atoms, and the temperature of different velocity layers. In this chapter, we use dilute local thermodynamic equilibrium `dilute-lte` to calculate the atomic level population and nebular local thermodynamic equilibrium `nebular` to calculate the ionization fraction, and both equilibria are part of TARDIS [3]. Secondly, the program evaluates the transition probabilities of atomic energy levels. TARDIS generates energy packets (an ensemble of photons with same energy) at the photosphere which propagate through the SN ejecta, calculates the electron optical depth by path-integrating the electron density on the trajectory, and simulates the photon-atom interaction optical depth by randomly sampling the atomic transition probabilities. When the optical depth of an energy packet reaches 1 during electron scattering process, its energy and direction are reassigned following Compton scattering process. When the optical depth of an energy packet reaches 1 during line interaction, its energy and direction are reassigned following the transition probabilities of atomic lines. By collecting the emitted energy packets, the program then compares the resulting luminosity to the input luminosity and updates the photospheric temperature and the temperature throughout the ejecta. Our experience shows that the convergence of the temperature structure can be achieved in about 20 such iterations.

There are two line-interaction strategies, `downbranch` and `macroatom`, available in TARDIS. In `downbranch` [4], photo-excited atoms are allowed to reemit photon with same excited energy or other de-excitation transition energies, and the de-excitation channel is selected according to transition probabilities. Based on `downbranch`, `macroatom` [5] is a more sophisticated line-interaction strategy which allows upward and downward internal transitions with different probabilities for a photo-excited particle. Considering `macroatom` is closer to the physical reality, while the computational time difference between two line-interaction strategies is not much, we choose `macroatom` for all of our calculations.

2.2 The Initial Guess Model

Our goal is to generate a large number of models that cover the parameter space of observed SNe as much as possible. For this purpose, we need a baseline ejecta model of which we will perturb that ejecta parameters to account for ejecta diversity.

We restricted our studies to models at around 19 days after explosion, which correspond to the time around optical maximum. We used the delayed-detonation (DD) model `5p0z22d20 _20_27g` [6, 7], and the W7 model [8] as that for the ejecta density profiles to derive an initial guess model (IGM) for the ejecta structure (see Fig. 2.1). These initial models were chosen for their success in providing reasonable fits to a number of supernovae in previous studies [7]. The exact details of these models are not important as they only serve as a starting point for the ejecta structures and will be modified later to generate the spectral library for further analyses.

The IGM was derived by comparing model spectral shapes with the data of two well-observed SNe: SN 2011fe and SN 2005cf. The spectrum of SN 2011fe was acquired at UT 2011−9−10 09 : 22 : 00 (0.39 days after B-band maximum luminosity) by HST/STIS [9]. The SN 2005cf spectrum was acquired at UT 2005−6−12 00 : 00 : 00 (the B-band maximum luminosity) from SSO-2.3m/DBS [10]. The original chemical profiles of (DD) model `5p0z22d20_20_27g` is not optimized for SN 2011fe and SN 2005cf. The density and chemical structure of model `5p0z22d20_20_27g` was adjusted manually to match the strongest observed spectral features of SN 2011fe and SN 2005cf. Major changes to the elemental abundances were introduced when the profiles of strong lines such as Si II 6,355 Å were examined closely. Increasing iron-group elements of DD `5p0z22d20_20_27g` improved significantly the spectral match in the UV to these two SNe.

The observed spectra were found to be well fitted by a model with the photospheric velocity being $v_{ph} = 7{,}300$ km/s and the integrated luminosity between 6,500 Å ∼ 7,500 Å being $10^{8.52}$ times solar luminosity. Note that to ensure the convergence of the temperature structure, the temperature profiles were calculated with 40 iterations using the default temperature convergence parameters (type:damped. damping_constant:1. threshold:0.05. fraction:0.8. hold_iterations:3. t_inner_damping_constant:1.). The default iteration in TARDIS is 20, which is usually sufficient to reach temperature convergence.

The temperature and density profile of the IGM model that fits major spectral features of SN 2011ef and SN 2005cf are shown in Fig. 2.1a, b, respectively. The chemical profiles of DD `5p0z22d20_20_27g` and the IGM are shown in Fig. 2.1c, d, respectively. The simulated spectra of the IGM are shown in Fig. 2.1d where we also show the spectra computed using DD `5p0z22d20_20_27g` for comparison. Note that in Fig. 2.1d, the flux levels of the models and the data were arbitrarily scaled to match the spectral features. The fits to the observed spectra of SN 2011fe and SN 2005cf across most spectral lines and UV continua are apparently better for the IGM model than for the original DD `5p0z22d20_20_27g` model. Note that we did not attempt to construct a quantitative model of the elemental structure at this stage.

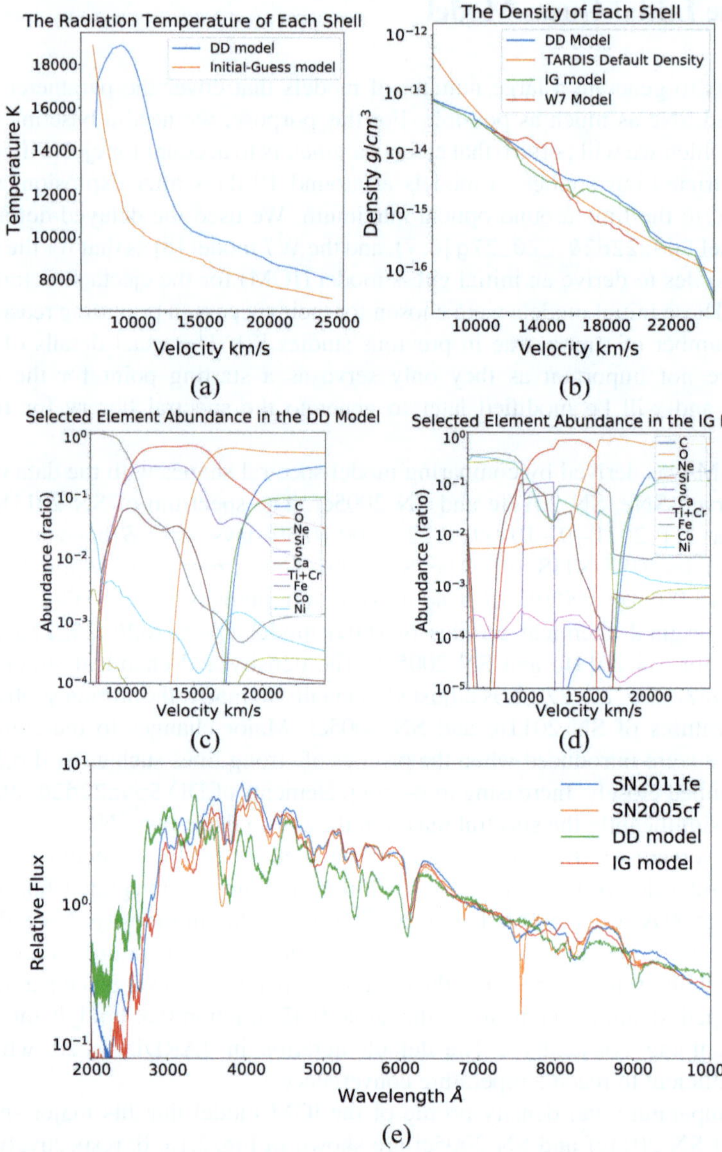

Fig. 2.1 **(a) Upper left:** The temperature in each shell of the DDT model 5p0z22d20 _20_27g and the IGM calculated with TARDIS. **(b) Upper right:** The density profile of the DD model, IGM, W7 model, and TARDIS' default 7-order approximated W7 model ($\rho = 3 \times 10^{29} \times v^{-7} \times (t/0.000231481)^{-3}\ \mathrm{g \cdot cm^{-3}}$, with t in days and v in km/s). All of the densities are normal SN Ia at 19 days after explosion. **(c) Middle left:** The elemental abundances of the DD model. **(d) Middle right:** The elemental abundances of the IGM, the elemental abundance ratios are normalized to 1 in each shell. **(e) Lower panel:** The TARDIS synthesized spectra of the DD model and the IGM, and the observed SN 2011fe / SN 2005cf spectra

This manual step only modified the masses of a limited number of elements (listed in Sect. 2.3) with prominent spectral features to achieve a rough fit to the data.

2.3 The Spectral Model Library

For our deep learning neural network, the IGM was used as a baseline model that was perturbed to generate a library of model spectra. The parameter space was constructed to cover a large parameter space. It was impossible to build a complete model grid with varying elemental abundances at all velocity layers considering the large number of chemical elements involved. For the structure of the ejecta, we divided the ejecta into four distinct zones defined by different velocity boundaries. The velocity ranges of these four zones are 5,690–1,0000 km/s, 10,000–13,200 km/s, 13,200–17,000 km/s, and 17,000–24,000 km/s. The ejecta structure includes all 23 elements with atomic number from 6 to 28. With four velocity zones, this leads to 92 free parameters on the masses of the chemical elements in each zone. The other parameters to run TARDIS but are not included in MRNN training are the luminosity, date of explosion, and photospheric velocity.

The chemical compositions and densities were allowed to fluctuate independently within each velocity zone, but inside each zone the velocity dependence of the density of each element is scaled to that of the same element of the IGM. For elements other than the iron group, we used the following equation:

$$\rho_{ik} = \rho_i^{IGM} \times 3U_k, \tag{2.1}$$

where i denotes each different chemical elements, k denotes the four different zones with $k = 0, 1, 2, 3$ stands for the velocity shells between 5,690–10,000 km/s, 10,000–13,200 km/s, 13,200–17,000 km/s, and 17,000–24,000 km/s, respectively, and U_k is a random number drawn from a (0,1] uniform distribution. The locations of the velocity boundaries were chosen to approximately match major chemical layers in the IGM, and ρ_i^{IGM} is the density profile of element i in the IGM of shell i. The total density including all the elements is the sum over all elements. With Eq. (2.1), the masses of the elements in the four zones were artificially scaled from 0 to 300% relative to their respective values in the IGM.

Because Zone 1 contains mostly Fe, Co, and Ni that are sensitive to the location of the photosphere, we choose the variation of Fe, Co, and Ni in Zone 1 to obey

$$\rho_{ik} = \rho_i^{IGM} \times 3^{(1-2U_k)}, \tag{2.2}$$

where U_k is again a random number drawn from a uniform distribution between 0 and 1. Equation (2.2) samples the chemical structures around ρ_i^{IGM} more frequently than large deviations from it. Hereafter, the element mass ratios between the library models and the IGM are delineated as "multiplication factor."

Table 2.1 Photospheric velocity and explosion date

Days after explosion	16	17	18	19	20	21	22	23
Photosphere velocity (km/s)	8090	7850	7430	7100	6650	6290	6050	5690

Another input parameter of TARDIS is the time after explosion. For the current study, we restricted us to models around optical maximum. This parameter was drawn from a uniform distribution between 16 and 23. The location of the photosphere was defined by interpolating the values given in Table 2.1, with an additional random number from a uniform distribution between −120 km/s and 120 km/s to sample the range of possible variations of the location of the photosphere. The ranges of the location of the photosphere in Table 2.1 were derived using the IGM as input to TARDIS for which the corresponding photospheric velocities provide reasonable matches to the observed spectra of SN 2011fe between day 16 and day 23 after explosion. In this process, we found the photosphere recesses approximately 300 km/s per day, which is consistent to the results of previous model fits by others [9].

TARDIS also requires the luminosity of the SN as an input parameter. For the spectral library, the luminosities in the range of 6,500–7,500 Å of the different models observe a uniform distribution in log-space between $10^{8.64}$ and $10^{8.7}$ solar luminosity.

TARDIS assumes a predefined sharp photosphere and does not calculate the gamma-ray transport in the ejecta. The radial temperature profile is only approximately correct and in some regions may be severely incorrect. This caveat makes it difficult to constrain physical quantities derived from TARDIS based on first principle physics. The models are calculated independently at each epoch. This may introduce errors that make it difficult to compare the model and observed luminosities. However, we expect the spectral profiles to be governed by fundamental physics, and the model spectra bear the imprints of the atomic processes. Furthermore, some of these problems indigenous to TARDIS may be coped with by comparing a subset of TARDIS generated models with more sophisticated radiative transfer codes such as CMFGEN [11] and HYDRA [7] to identify systematic differences among these models. These systematic comparisons may lead to a large library of supernova models with more physical processes being taken into account, but the calculations of a large library of CMFGEN or HYDRA models are too computationally costly to be practical. We leave a comparative study of different models to future studies.

We used TARDIS' default temperature convergence strategy in calculating these models but with 15 temperature iterations (i.e., type:damped. damping_constant:1. threshold:0.05. fraction:0.8. hold_iterations:3. t_inner_damping_constant:1.). Additionally, in order to generate spectra with different signal-to-noise ratios, we set the number of Monte Carlo packets to vary from 1.5×10^5 to 2×10^6. The wavelength coverage of the model spectra was set to 2,000–10,000 Å.

2.3.1 Model Spectra Computation

The calculation of a single spectrum of the spectral library with about 10^7 energy packets takes approximately 0.5 to 2 CPU hours on our workstation which is mounted with two Intel Xeon E5-2650 v4 CPUs. After about one month's calculation utilizing 80 cores (two chips of Intel Xeon E5-2650 v4, and one chip of Intel Xeon E5-2650 v1), a total of 99510 spectra were generated with varying elemental abundances, photospheric velocities, explosion dates, and luminosities parameter space, as prescribed in the previous section.

2.3.2 Response of Spectral Profiles to the Variations of Input Abundances

The working hypothesis of this study is that the spectral profiles are sensitive to the input parameters, and there is predictive power of the TARDIS models, albeit the spectral profiles may respond to variations of chemical elements that may be very complicated. We demonstrate in Fig. 2.2 that this is indeed the case. To show the effect of varying chemical elements, a single element at a given layer was artificially altered, while all other parameters remained the same as in the reference model. Here the reference model was chosen from the spectral library that best matches SN 2011fe at day 0.4 (see Sect. 2.5). Figure 2.2 shows the results of artificially altering Fe and Co by 0, 2, 4, and 8 times with respect to a reference spectrum.

We see that the spectral patterns of Fe and Co are very different. The spectra vary strongly the UV wavelength and create spectral "wiggles" that are characteristic to the input chemical elements. The features generated by elements in different velocity zones are also distinctively different.

2.4 The Multi-residual Connected CNN Model

In this section, we apply the multi-residual convolutional neural network (MRNN) [12] to the spectral library synthesized through the procedures outlined in Sect. 2.1 and train deep learning models to infer the ejecta structure from the synthesized spectral library.

2.4.1 Model Data Preprocessing

The spectra generated by TARDIS are in units of $erg/s/\text{Å}$, which represents the luminosities of the SNe. They have a typical value of $\sim 10^{38}$ around optical

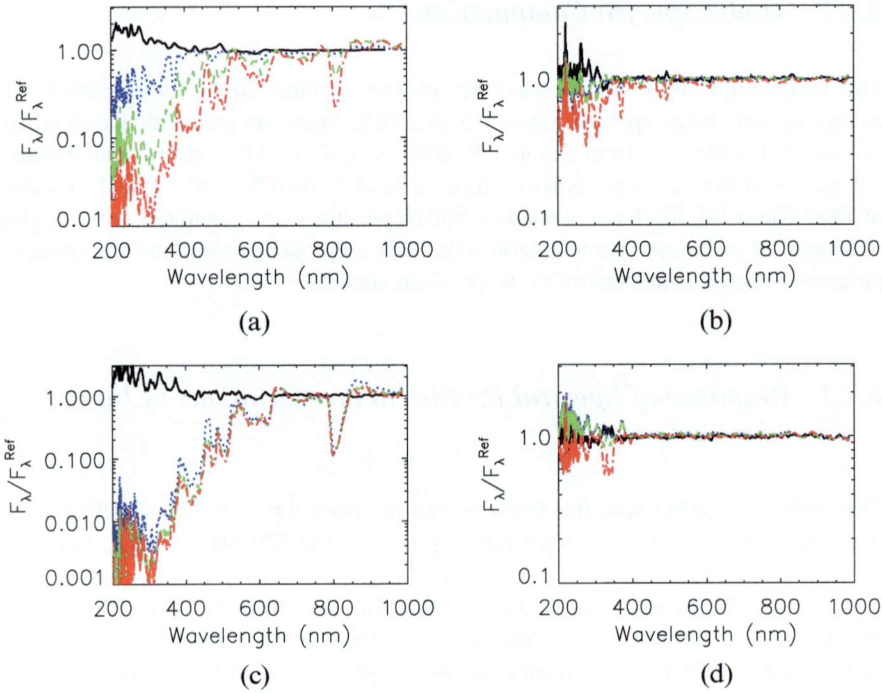

Fig. 2.2 From upper left clockwise, the panels show the effect on spectral profiles by only varying Fe in Zone 2 **(a)**, Fe in Zone 3 **(b)**, Co in Zone 2 **(c)**, and Co in Zone 3 **(d)**. The spectra were normalized by a reference spectrum which provides a good fit to the observed spectrum of SN 2011fe at day 0.4. Note that solid black lines with a scaling factor of 0 are models equivalent of removing that component from the ejecta. The dotted blue lines, dash green lines, and dash-dotted red lines are for models with that component enhanced by factors of 2, 4, and 8, respectively

maximum. This study explores the models in relative flux scale as TARDIS is more reliable in modeling spectral features than absolute luminosity. The flux of each spectrum is normalized by dividing its average flux between 6500 and 7500 Å. Only the overall spectral shapes are of importance here. The absolute level of the flux is ignored throughout the neural network analyses. A discussion on the correlation between spectral features and luminosity will be given after the establishment of the neural network (see Sect. 2.6.3).

Moreover, to account for the problems related to TARDIS Monte Carlo noise when comparing to observational data, we further employed two methods for data augmentation: Gaussian noisification and Savitzky-Golay filtering [13]. Gaussian noisification is done by applying a Gaussian noise for each wavelength bin following the formula $F_n(\lambda) = F_o(\lambda) \times (1+\mathcal{N}(0, F_o(\lambda))/F_o^{1/2}(5500)/S)$, where $\mathcal{N}(0, F_o)$ is a normal distribution with $\mu = 0$ and $\sigma^2 = F_o$, S is a measure of the signal-to-noise ratio at the reference wavelength 5500 Å for the spectra with noise added, and F_n and F_o are the flux of the noise added flux and the original TARDIS model flux, respectively. Savitzky-Golay filtering is achieved with smoothing windows randomly selected from (7, 9, 11, 13, 15, 17, 19, 21, 23, 25) and the order randomly chosen from (2, 3, 4, 5, 6). With this data augmentation strategy, we generated a new dataset that is 8 times larger than the original spectral library.

2.4.2 The Neural Network Structure

The SN spectra we aim to model are characterized by broad spectral lines due to the high velocity of the ejecta and the blending of atomic multiplets; models of SNe Ia require more complicated neural network structures to reach sufficient accuracy and sensitivity.

Between the input and output, a typical convolutional neural network (CNN) contains stacked convolutional layers above one or two fully connected layers, incorporated with pooling layers and activation layers. Starting from the input layer, the data propagate through the convolution layers via convolution with convolution cores and pass through the activation layer with nonlinear functions (i.e., rectified linear activation $f(x) = 0\ when\ x \leq 0; x\ when\ x > 0$). In the pooling layer, the dimensions of the data are reduced by binning adjacent data while preserving the maximum value (MaxPooling method) or the average value (AveragePooling method). The fully connected layer shares the same structure as normal neural networks, which calculates linear combinations using the weights and biases between every neurons in the adjacent layers.

The trainable parameters (mainly the weights and biases in the fully connected layer, and the convolution cores in the convolution layers) are randomly assigned at the outset and are updated in the training process through forward and backward propagation. Forward propagation calculates the output using the current neural network parameters and the input and then compares the network outputs (predictions) with the target values (the real values) under a loss function (i.e., mean squared error, MSE: $Loss = Mean(\hat{M}_{scaled} - \hat{M}_{predict})^2)$. As all the calculations in neural network are analytical, the gradients of all trainable parameters can be deduced. In the backpropagation process, the trainable parameters are updated by multiplying a predefined learning rate to their gradients (for a review, see [14]). Limited by computational efficiency, every forward-backward propagation iteration only avails a subset of the training dataset (batch), so a full review of the training dataset consists of several batches.

It is stated [15] and experimentally demonstrated (i.e., AlexNet [16], VGG16 [17]) that additional layers endow the neural network with exponential increase in agility. Nonetheless, a deeper network (a network with more layers) suffers from "gradient explosion" or "gradient diminish" problems: During the forward propagation calculations, an input signal may be consecutively multiplied with <1 or >1 numbers, which results in the signal turn to zero or infinity in machine accuracy and affects the calculation on gradients. As a consequence, [18] introduced batch normalization layers, which normalize the input batch in every dimensions to be average zero and standard derivative one. Additionally, [19] suggested to add the output from the previous convolution layers onto the current layer output and use this two convolutional layered structure as the building block for CNN structure. Such residual neural networks (ResNet) structure shows great accuracy and is easy to optimize even in 152-layered scenario [19]. Based on ResNet, [20] proposed densely connected neural networks (DNNs) with another signal shortcut method: Directly concatenate the output of all the previous layers as the input of the current layer, and insert low-dimension layers served as "bottle-necks" to break the cumulative dimension increase caused by the consecutive concatenations.

Consequently, we adopt the multi-residual neural network (MRNN) structure [12] for our studies. The structure of MRNN is relatively simple. Comparing to ResNet which allows signal shortcut in two layers, MRNN introduces signal shortcut in multiple layers by adding all the previous layers' (or blocks') outputs together to be the next layer's input. Such a structure is much simpler than DNN and requires less computation time to find a suitable network structure. Secondly, MRNN balances the training demands and the accuracy. According to the test on CIFAR-10 and CIFAR-100 image classification dataset [21], MRNN shows an equivalent performance compared to a much deeper ResNet structure [12] and better than all plain CNN structure available at that time. However, we notice our CNN and MRNN with similar depth consume comparable amount of CPU time to finish one epoch of training. Finally, we chose MRNN to probe the best-performance element prediction.

All the models were trained using `keras` [22] with `tensorflow` [23] as backend. In the preliminary training step to build up the neural network structure, we selected the iron abundance in Zone 3 and the TARDIS synthesized spectra between 2000 and 10,000 Å. We choose mean squared error (MSE) as the loss function and "adam" [24] as the optimizer. We adopted a two-step training scheme in order to achieve convergence, while avoiding overfitting (the model performs exceptionally well on the training dataset but fails on the testing dataset). In the first step, the learning rate is 3×10^{-5} and decays 10^{-8} per time step, the batch size is 4,000, and the training session jumps to the second step when there is no progress in the loss function of the testing dataset after 10 epochs. In the second step, the learning rate is 3×10^{-7} and decays 10^{-8} per time step, the batch size is 10,000, and the training session stops when there is no progress in the loss function of the testing dataset after 5 epochs.

The network architecture is shown in Fig. 2.3. The neural network structure for the absolute luminosities is similar, but with a different training schedule, details are discussed in Sect. 2.6.3. We also tried to train the neural network to predict the photospheric velocities and the date of maximum; however, we failed to use these two predictions to refit the observed spectra. For the spectral fits in Sect. 2.6, we adopted the explosion date from light curves and used grid search to find the best photospheric velocities instead.

We trained the neural network using spectral data with two different wavelength coverages. One set used the entire spectral wavelength range from 2,000 Å to 10,000 Å (WR-Full, hereafter), and the other only used the wavelength range from 3,000 Å

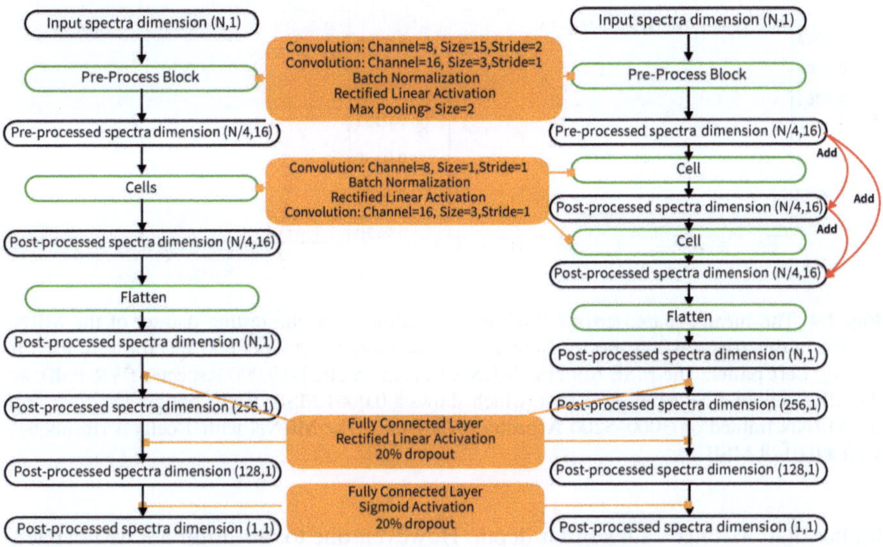

Fig. 2.3 Left panel: The CNN structure. The black squares are data tensors transferring in the network, and green squares are the "block" structures which contain several neural network layers, as is marked in the yellow squares. To notice, "cell" structures have the same size for its input and output and were used consecutively in MRNNs. For the 2,000–10,000 Å spectra, the sampling is 2000 pixels, and the tensors transform in neural network as (2000, 1) → (500, 16) → (8000, 1). As for the 3000–5200 Åspectra, the sampling is 719 pixels, and the tensors transform in neural network as (719, 1) → (180, 16) → (2880, 1). **Right panel:** An example of the MRNN with two cells. The outputs of each cell are added with all the inputs of previous cells to form the input of the next layer, as is marked by the red arrows. For the densely connected neural network, the "adding" processes are replaced with "Concatenate" process, so the data size of the intermediate, the processed, and the first fully connected layers are doubled

to 5,200 Å (WR-Blue, hereafter). These two sets of data are useful for applications to observational data with different spectral coverages.

In the next step, we chose the number of "cells" to be 1, 3, 5, 7, 9, 11, and 13 and explore the performances of MRNN. We randomly selected 20,000 spectra for training and reserved 2,666 spectra for testing. We found that MRNN with seven cells performs the best among all models, as is shown in Fig. 2.4. Increasing the number of cells actually deteriorates the fits.

We thus adopted the MRNN for the training on the chemical elements from 6 to 28 and the B-band absolute magnitudes. The chemical elements in the four different zones (plus the luminosity) were trained separately. In principle, the parameters can be trained jointly. However, the size of such a neural network could be too large for our computational capability, so we opted to choose a fast and robust neural network structure and applied it to all prediction tasks.

We have also tried the densely connected neural network (DNN) [20] on the 2000–10,000 Å models, by simply changing the "adding" process in the MRNN by "concatenation." The MSE when using three cells for calculation is 0.0080, which is

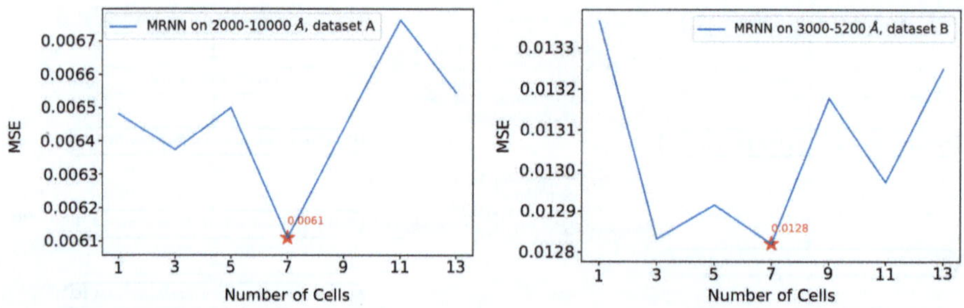

Fig. 2.4 The mean squared error (MSE) of Fe in Zone 3 on the testing dataset of the MRNN model. In this trial, 20,000 spectra were used in the training set and 2666 spectra were used for testing. **Left panel:** The MSE from the MRNN trained on 2000–10,000 Åspectra (WR-Full), and the MRNN with 7 cells performs best, which shows a 0.0061 MSE. **Right panel:** The MSE from the MRNN trained on 3000–5200 Åspectra (WR-Blue). The MRNN with 7 cells performs best, with a 0.0128 MSE

higher than MRNN with similar depth. However, due to our limited RAM capacity and the complexity in modifying DNN's cell structures, "bottle neck" [20] positions, and other hyper-parameters, we did not explore more of this network structure.

2.4.3 Target Parameters for Machine Learning

Our goal is to find the ejecta structures that best fit the observed spectral features of an SN Ia at around optical maximum. We choose the "multiplication factor" (as is mentioned in Sect. 2.1) in the zone of our concern as the neural network output. The "multiplication factor" was restricted to have a range from 0 to 3. Our experiments with the models and data indicate that this range is sufficient to ensure coverage of a broad range of SNe Ia. Then, we applied the following normalization strategy in order to constrain the output of the neural network into (0,1):

$$\hat{M}_{scaled} = \tanh\left(\frac{(\hat{m} - \mu(\hat{m}))}{\sigma(\hat{m})}\right)/2 + 0.5, \tag{2.3}$$

where $\hat{M}_{scaled}$ is the scaled elemental abundances of the neural network output, and μ and σ are the average and the standard derivative of the multiplication factor $\hat{m}$, respectively. This nonlinear normalization strategy allows the trained model to predict values outside the parameter space (extrapolation) within a small parameter range, but suppresses erratic values derived by the neural network; elemental abundances less than zero or approaching infinity are remapped to values close to 0 and 1, respectively.

2.4.4 Training Results

Not all the elements in the IGM are significantly influencing the spectra. As a first trial, we adopted a subset of model spectra and a simplistic neural network structure to probe the effect of various chemical elements. In this trial, 10,000 spectra were selected as the training set and 1829 spectra as testing set, and we chose the MRNN with one cell structure. All the elements from atomic number 6 (Carbon) to 28 (Nickel) in the four velocity zones are trained on the training set and verified on the testing set. By comparing the MSEs from the testing set, as well as the correlation between the neural network-predicted scaled elemental abundance (Eq. 2.3) and the corresponding value of the TARDIS input (Figs. 2.5 and 2.6), we found models with MSE larger than 0.1 to be poorly trained with little predictive power on elemental abundances. Consequently, we chose only the elements in zones with MSEs less than 0.1 for further training. There are 34 and 31 trainable chemical constituents located in the four velocity zones for train sets WR-Full and WR-Blue, respectively. The MSEs of these chemical constituents are shown in Tables 2.2 and 2.3 for train sets WR-Full and WR-Blue, respectively.

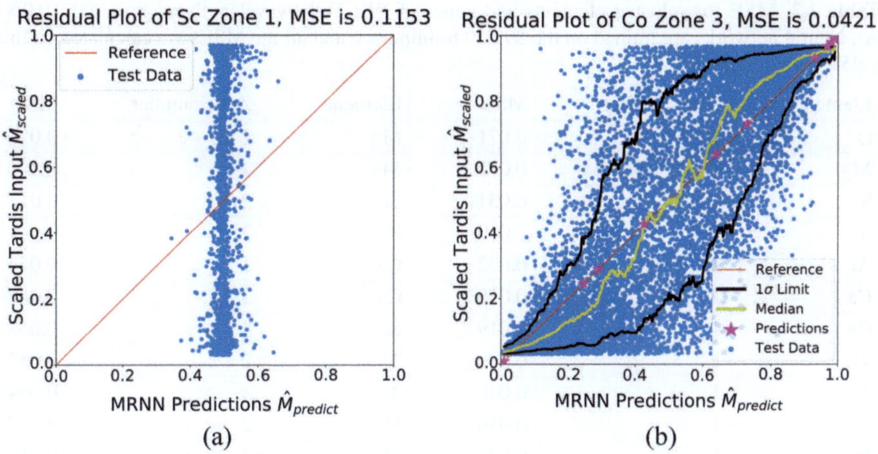

Fig. 2.5 The correlation between the scaled elemental abundances of MRNN predictions and TARDIS inputs. X-axis is the prediction from the MRNN model, and Y-axis is the input into TARDIS for spectral calculations. In both panels, the diagonal red lines indicate the ideal model predictions, and the blue dots were derived from the test dataset. Yellow and black lines show the median and $1 - \sigma$ error limit of the given prediction in the testing set. Predictions from the selected 11 SN Ia spectra with wavelength coverage from 2,000 to 10,000 Å (which are analyzed in Sect. 2.5) are shown as magenta stars. **(a) Left panel:** The correlation plot of Sc in Zone 1, MSE, is 0.1153. This model was trained on 10,000 spectra and tested on 1,829 spectra. The predicted values are close to 0.5, which violates the "real" value for spectral synthesis, indicating the poor performance of the neural network on this element zone. **(b) Right panel:** The plot showing the correlation of scaled elemental abundances of Co in Zone 3, with an MSE value of 0.042. This model is trained on 89,559 spectra and tested on 9,951 spectra

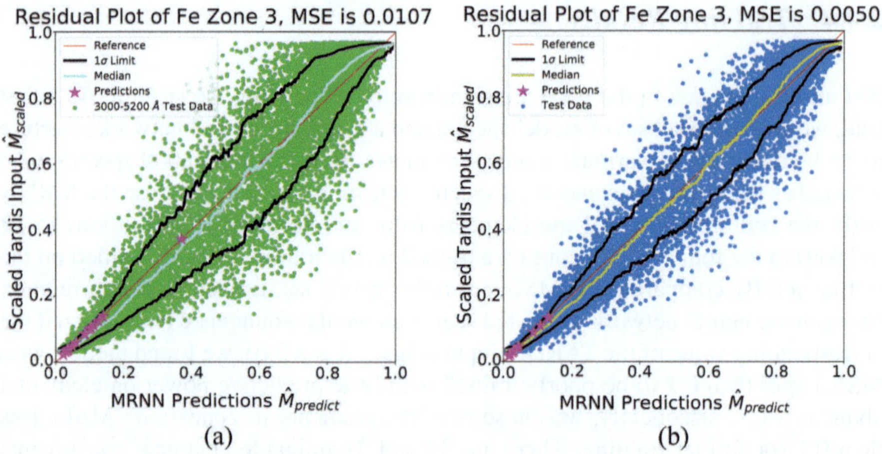

Fig. 2.6 Same as Fig. 2.5, but for Fe in Zone 3. **(a) Left Panel:** The correlation between scaled elemental abundances of MRNN predictions and TARDIS inputs for neural network trained for wavelength range 3,000–5,200 Å. The MSE is 0.0107. **(b) Right panel:** The same plot for wavelength range 2,000–10,000 Å. The MSE is 0.0050

Table 2.2 MSE for selected elements and zones of WR-Full (wavelength range 2,000–10,000 Å). Neural networks are trained on the 89,559 training set, and all the MSEs are calculated on the 9,951 testing dataset

Element	Zone number	MSE	Element	Zone number	MSE
O	4	0.071	Mg	2	0.010
Mg	3	0.075	Mg	4	0.040
Si	1	0.051	Si	2	0.013
Si	3	0.019	S	2	0.067
Ar	2	0.092	Ca	1	0.054
Ca	2	0.018	Ca	3	0.048
Ca	4	0.019	Sc	4	0.058
Ti	1	0.086	Ti	2	0.055
Ti	3	0.046	Ti	4	0.044
V	1	0.050	V	2	0.019
Mn	1	0.026	Mn	2	0.017
Fe	1	0.004	Fe	2	0.005
Fe	3	0.050	Fe	4	0.025
Co	1	0.088	Co	2	0.029
Co	3	0.042	Co	4	0.078
Ni	1	0.009	Ni	2	0.028
Ni	3	0.050	Ni	4	0.075

As discussed in Sect. 2.4.2, we adopted the MRNN with seven cells for elemental abundance estimation. The training dataset contains 89,559 spectra (90% of the total), and the testing dataset contains 9,951 spectra (10% of the total). For a typical

Table 2.3 MSE for selected elements and zones of WR-Blue (wavelength range 3,000–5,200 Å). Neural networks are trained on the 89,559 training set, and all the MSEs are calculated on the 9,951 testing dataset

Element	Zone number	MSE	Element	Zone number	MSE
Mg	2	0.070	Mg	3	0.086
Mg	4	0.051	Si	1	0.085
Si	2	0.058	Si	3	0.053
S	2	0.071	Ar	2	0.093
Ca	1	0.062	Ca	2	0.072
Ca	3	0.073	Ca	4	0.052
Sc	4	0.062	Ti	2	0.066
Ti	3	0.061	Ti	4	0.046
V	1	0.087	V	2	0.046
Mn	1	0.069	Mn	2	0.058
Fe	1	0.037	Fe	2	0.024
Fe	3	0.011	Fe	4	0.036
Co	1	0.030	Co	2	0.045
Co	3	0.046	Ni	1	0.039
Ni	2	0.056	Ni	3	0.051
Ni	4	0.080			

neural network, it takes approximately 1 hour to finish 200 training epochs on two Tesla P100 GPUs.

In Tables 2.2 and 2.3, we list the MSE on the testing dataset of the 34 and 31 selected elements and zones for two sets of neural networks built for WR-Full and WR-Blue, respectively. As an example, the scaled elemental abundances of the MRNN prediction and those of the TARDIS input of Fe in Zone 3 are shown in Fig. 2.6. The neural network-predicted chemical abundances clearly correlates with the values used to generate the model spectra. The neural network using the full 2000–10,000 Å wavelength coverage clearly outperforms the ones with partial coverage from 3000 to 5200Å. The fact that the relation between the RNN prediction and the original TARDIS input elemental mass is approximately given by $\hat{M} \approx \hat{m}$ in all cases for which we establish a correlation that indicates that the training dataset yields results that are consistent with the testing dataset. This ensures that overfitting is not severely affecting the MRNN we have constructed. However, overfitting does appear to be an issue for elements that have weak spectral lines and where the correlation between the MRNN predicted and the original TARDIS input values is weak. This can be seen from Fig. 2.5 where the predicted mass of Sc is biased toward the mean value of true values of TARDIS input (~0.5). For Co in Zone 3, the correlation can be detected, but the predicted values are biased toward values higher and lower than the true TARDIS inputs for input values close to 0 and 1, respectively. Such bias is much weaker when the correlations are strong, such as as shown in Fig. 2.6 for Fe in Zone 3. This bias can in principle be corrected by using the median values of the Scaled TARDIS Input $\hat{M}$ to estimate the original model input.

In the current study, we did not correct this bias. Instead, we used the original MRNN predictions directly as the estimates of the mass of the input chemical elements but set the confidence levels of the estimates according to the correlations that can be derived from the test dataset. We did not attempt to estimate the confidence intervals using Bayesian statistics based on the Markov Chain Monte Carlo algorithm (for example, see [25]), due primarily to limited processing power of our computers. Instead, we estimated the $1-\sigma$ error by using the testing set itself. For each input model abundances such as those shown as examples in the vertical axis of Fig. 2.6, we derive the $1-\sigma$ confidence levels based on the dispersion of the predicted values (the horizontal axis). To be more specific, we collected all the data in the test dataset that agree with the MRNN predictions to within $[-0.02, +0.02]$ to build a histogram of the scaled TARDIS input mass, and used the 15.8%, 50% (median), and 84.1% values of the histogram to compute the $1-\sigma$ error and the median value. The $1-\sigma$ errors of Fe in Zone 3 are over-plotted in Fig. 2.6 as an example. The errors for other elements in various zones are available in the online material.

Assuming that the TARDIS model spectra capture major spectral features of observed SN Ia, we may apply the neural network trained by theoretical models straightforwardly to observational data. When an observed spectrum is inserted into the well-trained MRNN, a set of chemical abundances and their associated uncertainties can be derived. Unfortunately, we do not have a large enough library of observed spectra to train the link between observational data and theoretical models. The consistency and reliability of the derived abundances can only be evaluated by comparing the results for different supernovae and their spectral time sequences.

Alternatively, we may consider the chemical elements determined by feeding the observed spectra to the MRNN as empirical parameters that yields optimal TARDIS fits to the observed spectra. These parameters do not necessarily serve as true estimates of the elemental abundances of the supernovae, but nonetheless, they can be used as model-based empirical parameters to analyze the properties of SN Ia.

Note that the neural network approach is different from a simple observation-to-theory spectral fit. The neural network model aims to match spectral features that are associated with differential chemical structure variations. Each chemical element in a certain zone is trained separately, and the neural networks thus trained are most sensitive to the changes of the spectral features related to this particular chemical element throughout the spectral range under study. Even though the spectral features of various elements are highly blended, the differential changes of spectral features due to varying chemical abundances are still detectable by the neural network we have constructed. Furthermore, the theoretical models may have intrinsic shortcomings and do not include all the essential physics. The assumption of a sharp photosphere, for example, cannot be correct in a strict sense. The lack of time dependence of the radiative transfer may also limit the precision of the theoretical models. However, our approach to elemental abundance may be less sensitive to these problems by construction although the current study cannot establish a quantitative assessment of the uncertainties caused by approximations intrinsic to TARDIS models.

Moreover, the neural network allows for the theoretical luminosity to be derived once the chemical structure of the ejecta is fully constrained. From the MRNN, the latter can be determined by spectral features without the need to know the absolute level of the spectral fluxes. When the global spectral profiles between observations and models are matched, the luminosity (or the temperature of the photosphere) is then completely constrained theoretically. This provides a theoretical luminosity of the supernova under study that is independent of the flux calibration of the spectra.

2.5 Applications of the Neural Network to Observational Data

Assuming the AIAI method constructed using MRNN in Sect. 2.4 to be correct, we may apply it to observed spectra as a sanity check of the method. The parameter space describing the ejecta structure is so large that our spectral library only covers it only sparsely. The chance of having a perfect fit to any specific observation with spectra already in the library is low. The model spectra were recalculated by using TARDIS with the ejecta structures determined from MRNN. To derive the optimal TARDIS models, we also allowed the photospheric positions and the luminosities of the supernovae to vary.

2.5.1 Applications to SNe with Wavelength Coverage **2000–10,000 Å**

There are six well-observed SNe with a wavelength coverage from 2000 to 10,000 Å. They are shown in Table 2.4. The UV data were all acquired using *HST*. These data are rebinned to 1 Å/pixel and are normalized by dividing their respective average flux between 6,500 and 7,500 Å, similar to what was done for TARDIS model spectra during neural network training.

2.5.1.1 SN 2011fe

SN 2011fe was detected on Aug. 24, 2011 in M101, located at a distance of approximately 6.4 Mpc [36]. Its luminosity decline in the B-band within 15 days after maximum brightness (hereafter, $\Delta M_{B,15}$) was 1.12 ± 0.05 mag [27]. Based on optical and radio observations, SN 2011fe is not heavily affected by any circumstellar material (CSM) or galactic dust extinction [37, 38]. Consequently, we did not introduce extra host galaxy extinction correction for SN 2011fe spectra.

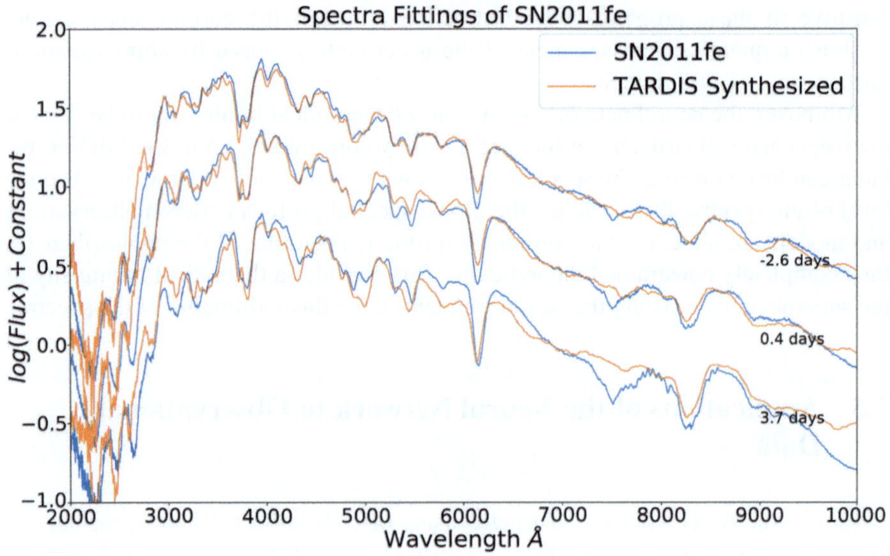

Fig. 2.7 The observed spectra (blue line) and the TARDIS synthesized spectra (orange line) of SN 2011fe at −2.6 days (upper), 0.4 day (middle), and 3.1 days (lower). The TARDIS synthetic spectra were filtered by a Savitzky-Golay filter with window =9, order =1 to remove Monte Carlo noise. The spectra are in logarithmic scale and are arbitrarily shifted in the vertical axis for the different dates

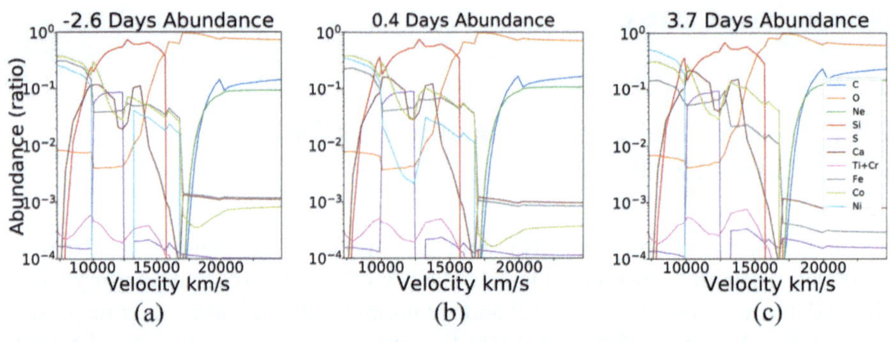

Fig. 2.8 The elemental abundances predicted by the neural networks from SN 2011fe at −2.6 days (**a**, left panel), 0.4 days (**b**, middle panel), and 3.7 days (**c**, right panel). These elemental abundances and the densities in Fig. 2.9 were used for the synthetic spectra shown in Fig. 2.7

We chose the *HST* spectra at −2.6 days, 0.4 days, and 3.7 days relative to the *B*-band maximum date for the elemental abundance calculations and spectral fits. The results are shown in Fig. 2.7.

The temperature profiles of the TARDIS models of SN 2011fe in the three spectroscopically observed phases are shown in Fig. 2.9. The three supernova structures shown in Fig. 2.8 are predicted by the neural network individually for each epoch, and the density structures may not strictly observe homologous expansion.

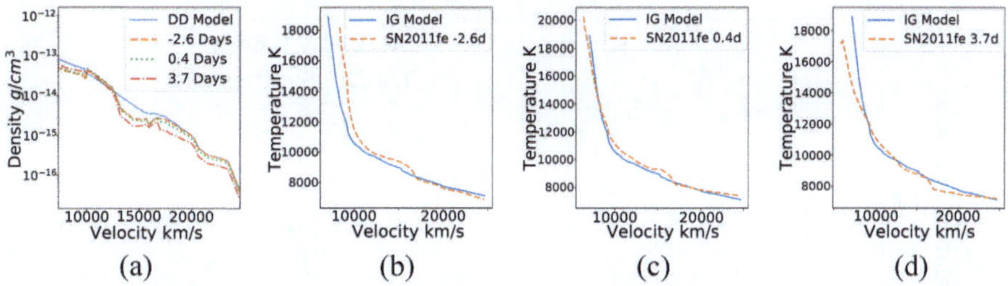

Fig. 2.9 **From left to right:** **(a)** The density structures of SN 2011fe at −2.6 days (orange line), 0.4 days (cyan line), and 3.7 days (red line) as predicted by the neural networks. The DD model is shown in blue line for comparison. **(b)** The temperature structure for SN 2011fe at −2.6 days. **(c)** The temperature structure for SN 2011fe at 0.4 days. **(d)** The temperature structure for SN2011fe at 3.7 days. Note that for convenience of comparisons we have converted all density profiles to 19 days after explosion assuming homogeneous expansion, using the relation $\rho \propto t^{-3}$. The temperature profiles for the IGM are shown in blue lines for comparison, and MRNN fits are shown as red lines

Table 2.4 Extinction and stretch of SNe Ia with 2000–10,000 Åspectra

SN name	Redshift	MW E(B-V)	Host E(B-V)	Stretch	Reference
SN2011fe	0.000804	0	0	1.062 ± 0.005	[26, 27]
SN2011by	0.002843	0.013	0.039	0.93 ± 0.02	[28–30]
SN2013dy	0.00389	0.135	0.206	1.098 ± 0.008	[31, 32]
SN2015F	0.0049	0.175	0.035	0.912 ± 0.005	[26]
SN2011iv	0.006494	0	0	0.830 ± 0.007	[33]
ASASSN-14lp	0.0051	0.33	0.021	1.101 ± 0.004	[34, 35]

However, it is encouraging to see that the density structures in Fig. 2.9a show adequate similarity with each other, and the corrections compared to the DD model appear to be consistent at different velocity layers; this cross-validates the prediction of the neural networks.

2.5.1.2 SN 2011iv

The transitional supernova (between type Ia normal and type Ia-91bg like) SN 2011iv was located in NGC 1404 with $\Delta M_{B,15} = 1.69 \pm 0.05$ [39]. According to [33], this supernova had negligible dust extinction effect along the line of sight so we did not apply any extinction corrections (see Table 2.4).

In the elemental prediction and spectral fitting process, we adopted the combined spectra of *HST* and the Magellan telescope at 0.6 days after the *B*-band maximum date [33]. The results are shown in Fig. 2.10.

The model spectrum agrees reasonably well with the observed spectrum across major spectral features in the optical. The disagreement across OI 7300 Å line is obvious. This is likely due to an insufficient amount of oxygen even though its

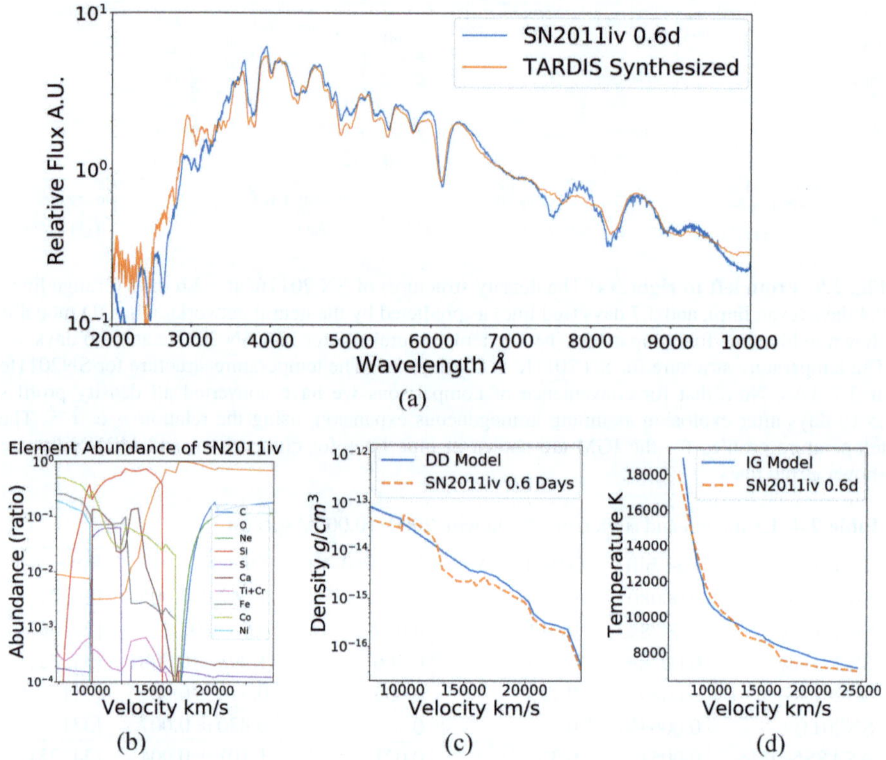

Fig. 2.10 **(a) Upper panel**: The observed spectrum (blue line) and the TARDIS synthetic spectrum (orange line) of SN 2011iv at 0.6 days after B-maximum. **(b) Lower left**: The elemental abundances of SN 2011iv predicted from neural networks. **(c) Lower middle**: Density structure of SN 2011iv predicted from neural networks (orange line) and the DD model density structure for comparison (blue line). Both densities are converted to that of day 19 using $\rho \propto t^{-3}$ relation. **(c) Lower right**: The temperature structure for SN2011iv spectral fit (orange line) and the IGM temperature structure for comparison (blue line)

abundance is three times that of the IGM. A similar problem may also be seen in Fig. 2.7 for SN 2011fe. We also see that the spectral features below 3,000 Å are poorly fit, suggesting again the elemental structure may not have appropriately covered this transition Type SN Ia. We will improve these fits in future studies with the construction of a larger spectral library.

2.5.1.3 SN 2011by

SN 2011by in NGC 3972 had $\Delta M_{B,15} = 1.14 \pm 0.03$ [40]. It had remarkably similar optical spectra and light curves to those of SN 2011fe, and these were identified as optical "twins" [29]. The only prominent difference is that SN 2011fe is significantly more luminous in the UV ($1600 < \lambda < 2500$ Å) than SN 2011by before and around peak brightness [30]. However, based on the distance deduced

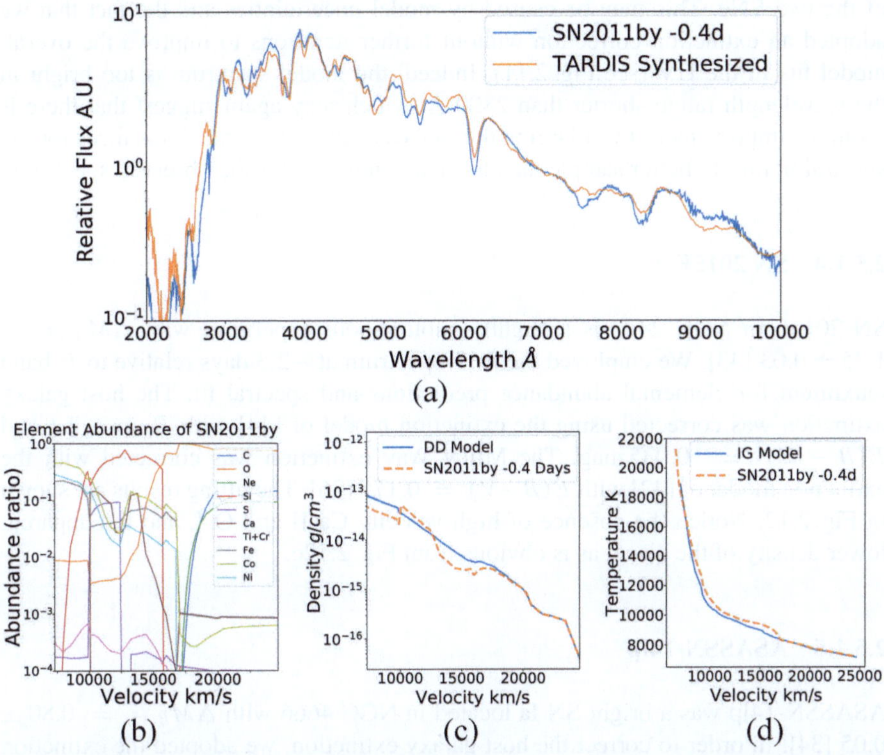

Fig. 2.11 **(a) Upper panel**: The extinction-corrected observed spectra (blue line) and the TARDIS synthesized spectra (orange line) of SN2011by at −0.4 days. **(b) Lower left**: The element abundance of SN2011by predicted from neural networks. **(c) Lower middle**: Density structure of SN2011by predicted from neural networks (orange line) and the DD model density structure as comparison. Both the densities are converted into 19 days scenario using $\rho \propto t^{-3}$ relation. **(d) Lower right**: The temperature structure for SN2011by spectral fit (orange line) and the IG model temperature structure as comparison (blue line)

from Cepheid variables in NGC 3972, SN 2011by had an absolute magnitude about 0.335 ± 0.069 mags dimmer than that of SN 2011fe [30]. This apparent magnitude difference can be a concern for supernova cosmology as its origins are unknown and is thus difficult to be corrected.

As in [30], we adopted the extinction model of [41] with $R_v = 3.1$ and $E(B-V) = 0.039$ to correct the host galaxy extinction, and Milky Way extinction models of [42] with $E(B-V) = 0.013$ to correct for Milky Way extinction. The *HST* spectrum of SN 2011by at 0.4 days before the B-band maximum was used to derive the elemental abundances using MRNN. The extinction-corrected observed spectrum and the corresponding synthetic spectrum are shown in Fig. 2.11.

It is worth noticing that the absolute luminosity of SN 2011by found by the MRNN process is slightly higher than that of SN 2011fe. Our MRNN spectral fits thus do not provide a theoretical explanation to the apparent luminosity difference of the two SNe. This may be caused by model uncertainties and the fact that we adopted an extinction correction without further iterations to improve the overall model fits in the UV (see Fig. 2.11). Indeed, the model spectrum is too bright in the wavelength range shorter than 2800 Å, which may again suggest that there is room for improvement by enlarging the ranges of the elemental abundances for the spectral library to better sample the physical conditions for the observed spectrum.

2.5.1.4 SN 2015F

SN 2015F in NGC 2442 is a slightly subluminous supernova with $\Delta M_{B,15} = 1.35 \pm 0.03$ [43]. We employed the *HST* spectrum at −2.3 days relative to B-band maximum for elemental abundance predictions and spectral fit. The host galaxy extinction was corrected using the extinction model of [44] with $R_v = 3.1$ and $E(B - V) = 0.035$ mag. The Milky Way extinction was corrected with the extinction model of [42] with $E(B-V) = 0.175$ [26]. The fitting results are shown in Fig. 2.12. Notice the absence of high-velocity Ca II and O I, and the apparent lower density of the ejecta as is obvious from Fig. 2.12c.

2.5.1.5 ASASSN-14lp

ASASSN-14lp was a bright SN Ia located in NGC 4666 with $\Delta M_{B,15} = 0.80 \pm 0.05$ [34]. In order to correct the host galaxy extinction, we adopted the extinction model of [44] with $R_v = 3.1$ and $E(B - V) = 0.33$ mag [34]. For Milky Way extinction, we adopt the extinction model of [42] with $E(B - V) = 0.021$ mag.

We used the *HST* spectrum at −4.4 days from B-band maximum for both the elemental prediction and spectral fits. The results are shown in Fig. 2.13.

The ejecta show enhanced density at velocity above 17,500 km/s. The Ca II H and K and IR triplet are clearly detected and highly blueshifted. The temperature profile shown in Fig. 2.13d is higher than that of the IGM throughout the ejecta. This is consistent with typically expected for SNe with slow decline rates.

2.5.1.6 SN 2013dy

SN 2013dy was located in NGC 7250 with $\Delta M_{B,15} = 0.90 \pm 0.03$ [31]. Like ASASSN-14lp, SN 2013dy is of the group with slow decline rates. We adopted the extinction model of [44] with $R_v = 3.1$ and $E(B-V) = 0.206$ mag to correct the host galaxy extinction, and the extinction model of [42] with $E(B - V) = 0.135$ mag for Milky Way reddening correction. The extinction parameter $E(B - V)$ was taken from [45].

The *HST* spectra at − 3.1 days, − 1.1 days, 0.9 days, and 3.9 days were used for modeling. The results are shown in Fig. 2.14.

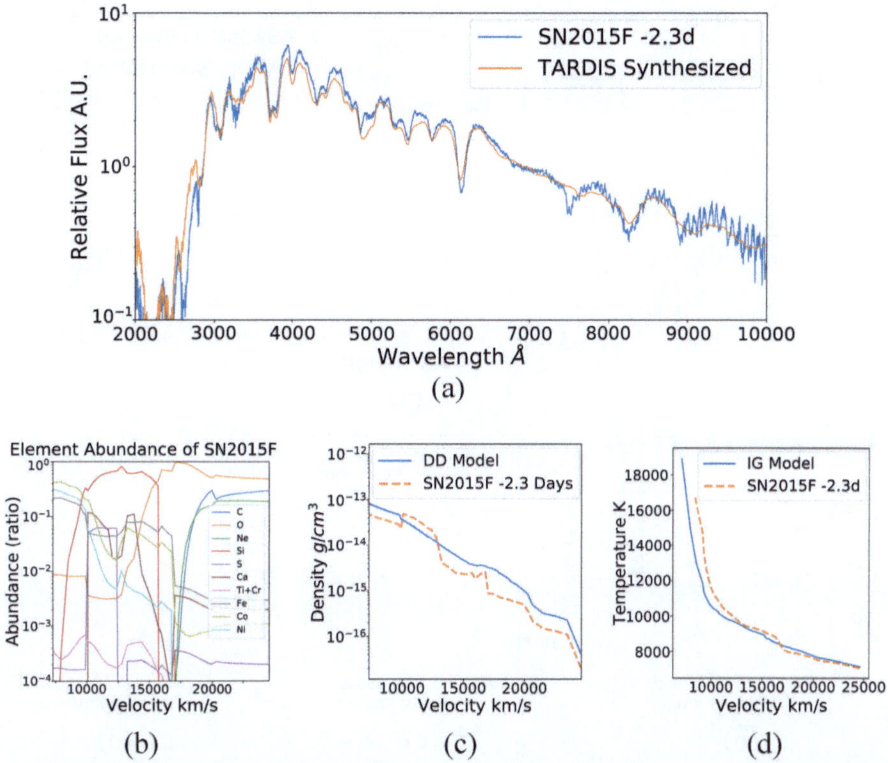

Fig. 2.12 **(a) Upper panel**: The extinction-corrected observed spectra (blue line) and the TARDIS synthesized spectra (orange line) of SN 2015F at −2.6 days. **(b) Lower left**: The elemental abundances of SN 2015F predicted by MRNN. **(c) Lower middle**: Density structure of SN 2015F predicted from MRNN (orange line) and the DD model density structure for comparison. **(d) Lower right**: The temperature structure for SN 2015F spectral fit (orange line) and the IGM temperature structure for comparison (blue line)

The model for SN 2013dy again shows consistently higher density at velocities above 17,500 km/s and higher temperature before optical maximum than that of the IGM. The O I feature at 7500 Å is again poorly fit. It implies that the oxygen abundances are too low for the spectra in the spectral library (Fig. 2.15).

2.5.1.7 Model and Observed Luminosities

Although the MRNN was trained in arbitrary flux scale which relies only on the profiles of spectral features, the final TARDIS spectra computed with the ejecta structure from MRNN do provide theoretical estimates of the luminosities of the SNe. The luminosity may be incorrect due to limitations of TARDIS models, but they may still be quantities correlated with the intrinsic luminosity of the

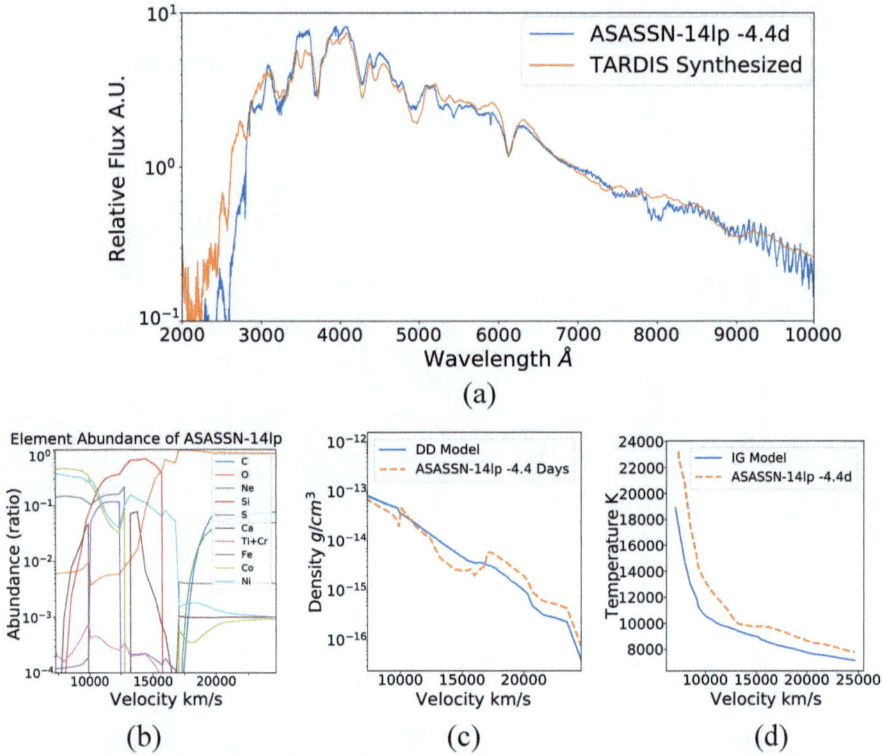

Fig. 2.13 **(a) Upper panel**: The extinction-corrected observed spectrum (blue line) and the TARDIS synthesized spectrum (orange line) of ASASSN-14lp at −4.4 days. **(b) Lower left**: The elemental abundances of ASASSN-14lp predicted by MRNN. **(c) Lower middle**: Density structure of ASASSN-14lp predicted by MRNN (orange line) and the DD model density structure for comparison. Both densities are converted into 19 days from B-maximum using the $\rho \propto t^{-3}$ relation. **(d) Lower right**: The temperature structure for ASASSN-14lp (orange line) and the IGM temperature structure for comparison (blue line)

supernovae. It is still interesting to see how they are correlated with the observed luminosities of SNe Ia. To compare the luminosities derived from the best match TARDIS spectra with those from more conventional methods, we employed the `SiFTO` [46] program to calculate the stretch factors of SN2011fe, SN2011iv, SN2011by, SN2015F, ASASSN-14lp, and SN2013dy. The resulting stretch values are listed in Table 2.4. The stretch values are known to be correlated to the SN luminosity [47, 48].

The model and true luminosities seem to be correlated but with an offset of about 0.7 mag. This difference is likely to be related to TARDIS model assumptions.

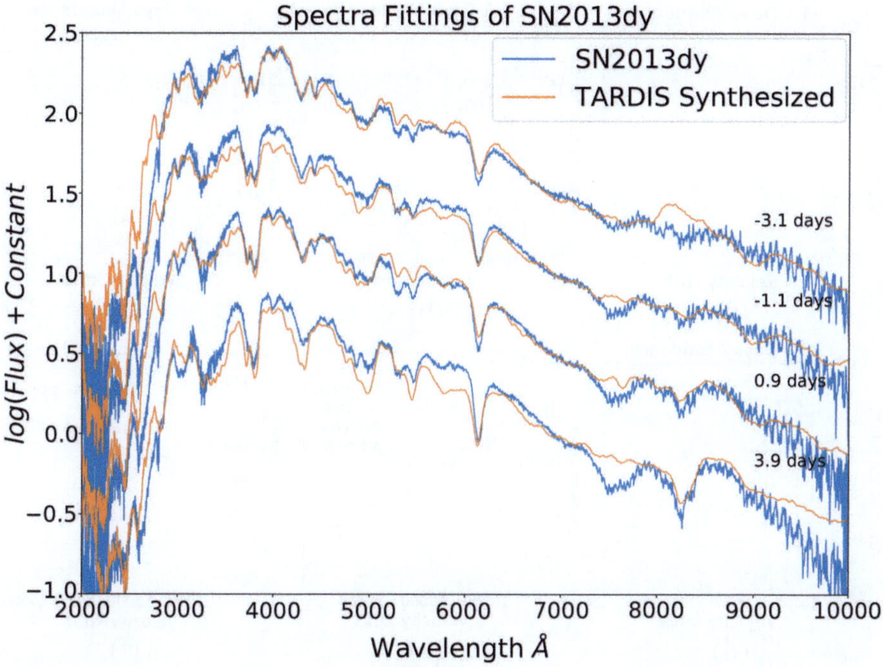

Fig. 2.14 The extinction-corrected observed spectra (blue lines) and the TARDIS synthetic spectra (orange lines) of SN 2013dy at −3.1 days, −1.1 days, 0.9 days, and 3.9 days. The fluxes are in logarithmic scale, and the spectra of different dates are arbitrarily offset by a constant

2.5.2 Application to SNe with Wavelength Coverage 3000–5200 Å

2.5.2.1 HST Spectra of 15 PTF Targets

Maguire et al. [28] presented 32 low-redshift ($0.001 < z < 0.08$) SN Ia spectra; the UV observations were obtained with *HST* using the Space Telescope Imaging Spectrograph (STIS). The wavelength coverage of these data is 3,000–5,200 Å. Meanwhile, the photometric observations were obtained by Palomar Transient Factory (PTF) [49, 50]. There are no published light curves of these targets yet. As our models are built only for data close to optical maximum, we selected only spectra taken between −3 and 4 days relative to the date of B-band maximum.

2.5.2.2 Dust Extinction Correction

To model the observed supernova luminosity, we need to properly treat the dust reddening by the Milky Way and the host galaxies.

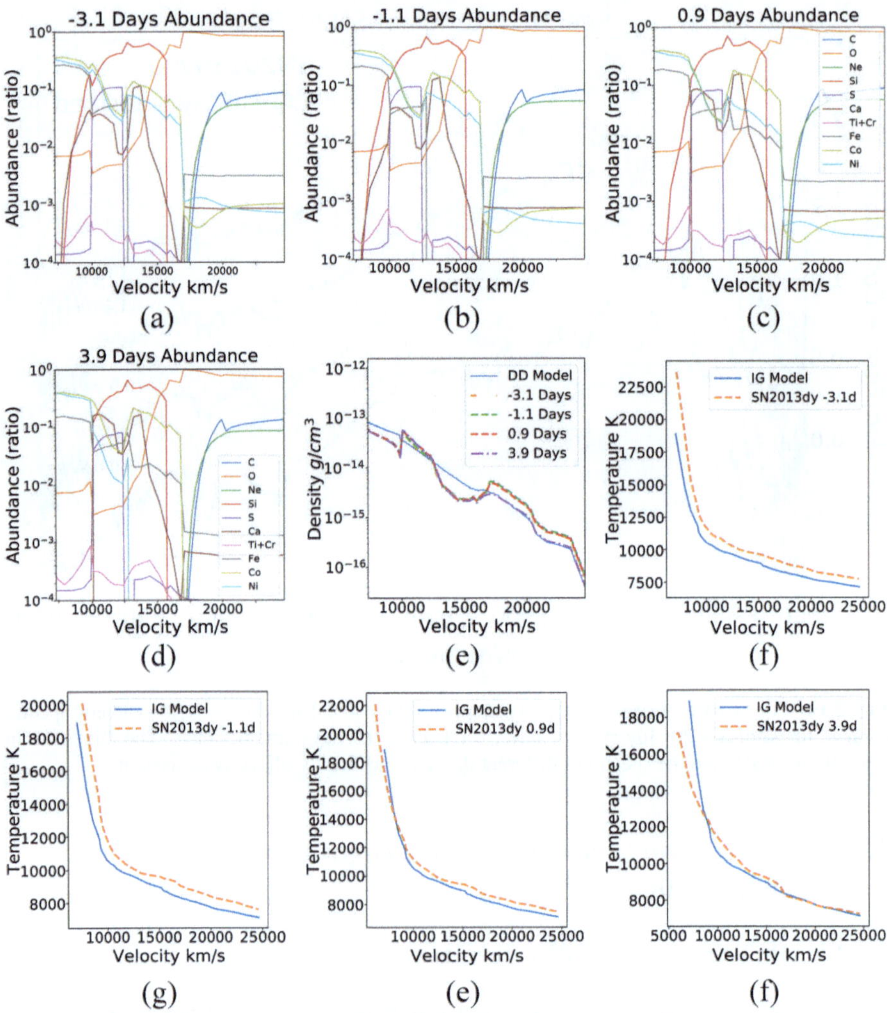

Fig. 2.15 **(a) Top left, (b) Top center, (c) Top right, and (d) Middle left**: Elemental abundance structure of SN 2013dy predicted by neural networks from the spectra at −3.1, −1.1, 0.9, and 3.9 days after B-maximum. **(e) Middle center** The density structure of SN2013dy predicted from the spectra of −3.1, −1.1, 0.9, and 3.9 days. The density profiles are converted into the profile of Day 19 using the $\rho \propto t^{-3}$ relation. **(f) Middle right, (g) lower left, (h) lower center, and (g) lower right**: The temperature structure for SN2013dy spectrum of day −3.1, −1.1, 0.9, and 3.9, respectively

The extinction data of the 15 *HST* spectra in [28] are not available. In order to correct the reddening effect, we developed an iterative algorithm by searching for the best match between theoretical models and observations with reddening as one of the χ^2 minimization parameters, as shown in the flow chart in Fig. 2.16. The procedure starts with an initial guess of reddening index $E(B-V)$, which is applied

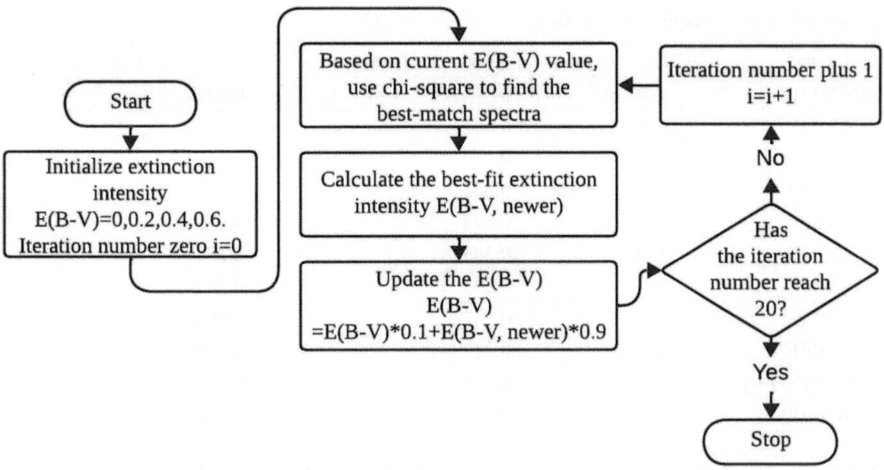

Fig. 2.16 An illustrative flow chart of the algorithm for dust reddening estimates (see Sect. 2.5.2.2 for details)

to the observed spectrum, it then calculates the RMS of the difference between the observed spectrum and all the models in the spectral library to find the best match, a new reddening index $E(B-V)_{newer}$ is calculated assuming the best-fit model is the true unreddened spectrum, the reddening index is updated using the formula $E(B-V) = 0.1 \times E(B-V) + E(B-V)_{newer} \times 0.9$, and the procedure repeats for 20 iterations to ensure convergence of the algorithm.

The algorithm was found to be sensitive to the initial value of $E(B-V)$, so we started the process using 0, 0.2, 0.4, and 0.6. With different extinction values and the resulting best-fit spectra, we compare the spectral fidelity and adopt the $E(B-V)$ value and spectral model with the least RMS difference between the observed and the model spectra reddened by the $E(B-V)$ value.

We also tested our extinction correction algorithm on the 11 SN spectra with wavelength coverage of 2000–10,000 Å and compare the derived extinction values with the extinction values in published literature. The results are shown in Table 2.5. We found our algorithm can reproduce low and intermediate extinctions, while for high extinction case $E(B-V) \sim 0.3$, the algorithm seems to underestimate the extinction intensity. As it is difficult to assess the precisions of flux calibrations of the observed spectra, we consider the values shown in Table 2.5 to be broadly in agreement.

In Table 2.6, we list the extinction values derived from this procedure for the 15 SNe under study.

To notice, several SNe show a density kink at velocities around 13,000 km/s. This is obviously an important issue as a density kink at such high velocity may be an important discriminator of explosion models of SNe Ia. The published DD models do not produce such a density kink. It was argued based on physical models that the density profile of delayed-detonation models may be quite robust, and is

Table 2.5 The extinction comparison

SN name	Phase (days)	$E(B-V)$ from papers	$E(B-V)$ from algorithm
SN2011fe	0.4	0	0.021
SN2011fe	−2.6	0	0
SN2011fe	3.7	0	0
SN2013dy	−3.1	0.341	0.222
SN2013dy	−1.1	0.341	0.253
SN2013dy	0.9	0.341	0.231
SN2013dy	3.9	0.341	0.253
SN2011iv	0.6	0	0.092
SN2015F	−2.3	0.21	0.129
ASASSN-14lp	−4.4	0.351	0.281
SN2011by	−0.4	0.052	0.018

Table 2.6 The *HST* observed SNe Ia from [28]. Only cosmological redshifts are given in [28], and the collected redshift data here are adopted from WISeREP

SN name	Phase (days)	Redshift	Stretch	$E(B-V)$
PTF09dlc	2.8	0.068	1.05 ± 0.03	0.21
PTF09dnl	1.3	0.019	1.05 ± 0.02	0.04
PTF09fox	2.6	0.0718	0.92 ± 0.04	0.08
PTF09foz	2.8	0.05	0.87 ± 0.06	0.15
PTF10bjs	1.9	0.0296	1.08 ± 0.02	0.02
PTF10hdv	3.3	0.054	1.05 ± 0.07	0.16
PTF10hmv	2.5	0.032	1.09 ± 0.01	0.02
PTF10icb	0.8	0.086	0.99 ± 0.03	0.13
PTF10mwb	−0.4	0.03	0.94 ± 0.03	0.02
PTF10pdf	2.2	0.0757	1.23 ± 0.03	0.23
PTF10qjq	3.5	0.0289	0.96 ± 0.02	0.14
PTF10tce	3.5	0.041	1.07 ± 0.02	0.23
PTF10ufj	2.7	0.07	0.95 ± 0.02	0.15
PTF10xyt	3.2	0.049	1.07 ± 0.04	0.24
SN2009le	0.3	0.017786	1.08 ± 0.01	0.17

insensitive to the details of the DD transition [51]. The kink can be results of ejecta-CSM interaction, or alternatively, it may agree with scenarios involving a detonation on the surface before the detonation at the center of the progenitor white dwarf [52]. However, the total mass of each velocity zone still suffers large errors as some elements such as oxygen and silicon are still poorly trained in the current version of the neural network. It may also be the results of the artificial division into four distinct zones. For the current study, we consider the reality of density kinks to be undetermined and leave further scrutiny to future studies with models using finer velocity zones.

2.6 Results and Discussions

2.6.1 Elemental Abundances and the Intrinsic Brightness

When applied to observational data, AIAI can be used to deduce the geometric structures of the ejecta. Even though the results are sensitive to physical approximations made in TARDIS, the results can nonetheless be used empirically to make a differential study between models and observations.

A few more clarifications of the uncertainties of the derived chemical abundances are necessary before carrying out such a study. Based on the testing dataset discussed in Sect. 2.4, we can estimate the $1-\sigma$ limits of the elemental abundances. Not all the predictions from the neural network are reliable due to the sparsity of our parameter space coverage in generating the dataset and the limitations of the sensitivity of the neural network. For example, some of the predictions of Co in Zone 3 are not in the region where the testing dataset has sufficient coverage, as are shown in the upper-right panel of Fig. 2.5b. When this situation occurs, we replace the predictions that go above or below the relevant $1-\sigma$ region with an upper limit or lower limit, respectively. Also, for simplicity, we have set the lower limits of the ejecta structure to be 7100 km/s when calculating the abundances of Zone 1 to avoid the effect introduced by a varying photospheric velocity.

The relation between elemental abundances and intrinsic luminosity of the supernovae is among the most interesting ones to explore. Taking the elemental masses derived from AIAI at their face values, we show in Fig. 2.17 the correlations between the stretch parameters and Fe, Co, and Ni masses all corrected to the values at 19 days after explosion.

Although we intend to leave a thorough analysis of these correlations to an upcoming paper, we identify immediately from Fig. 2.17 that the Fe, Co, and Ni masses in Zone 4 seem to be more strongly correlated to the stretch parameter than other parameters. Figure 2.17c also shows that the masses of Ni in Zones 2, 3, and 4 are all correlated to the stretch parameter. The mass derived from Zone 1 may not be reliable as the photosphere is located in Zone 1. The correlations that can be identified for Fe and Co in Zone 4 and Ni in Zones 2, 3, 4 are suggestive of the presence of radioactive materials at the surface of the supernova ejecta and are likely critical measures of the luminosities of SNe Ia.

It is noteworthy that the Ni mass in Zones 2, 3, and 4 varies by more than an order of magnitude. The masses of Fe and Co do not appear to share such behavior. No strong correlation is found between the masses of Fe and Co and the stretch factor either. This may be explained if a fraction of Fe and Co is nonradioactive and a significant fraction Ni in these zones is radioactive, or alternatively that the errors of the elemental masses derived from our neural network, being of the order of a factor of a few, are still to be large to reveal these correlations.

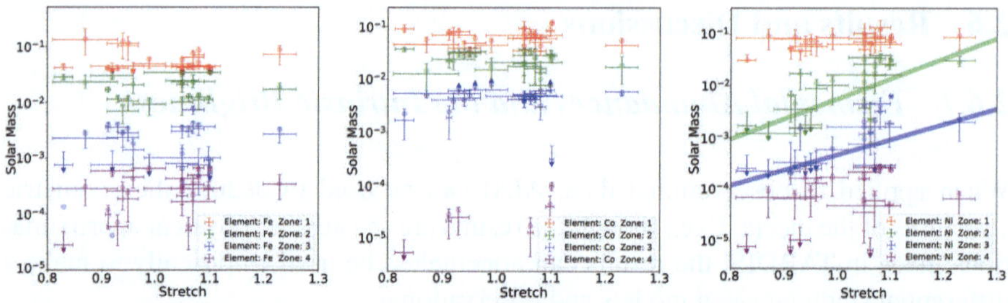

(a) (b) (c)

Fig. 2.17 The correlation between the stretch parameter and the mass of various chemical elements. The lower limits and the upper limits are marked with triangles, the median values are the crosses, and the predicted values are transparent circular dots. For simplicity, we choose the lower limit of the ejecta structures to be 7100km/s when calculating the element masses. **(a) Upper left**: The correlation between the stretch parameter and Fe mass in Zones 1, 2, 3, and 4. **(b) Upper right**: The correlation between the stretch parameter and Co mass in Zones 1, 2, 3, and 4. **(c) Lower left**: The correlation between stretch and Ni mass in Zones 1, 2, 3, and 4. For the neural network WR-Blue (wavelength range 3000–5200 Å), the Co mass in Zone 4 cannot be determined so we did not predict the Co mass in Zone 4 for the 15 spectra with such wavelength coverage. Notice that the Ni mass is corrected by its radioactive decay, and the predictions are for the Ni mass at the B-band maximum rather than the date of observations. We fitted the Ni mass in Zone 2 and Zone 3 with $M_{Ni,2} = 7.511 \times 10^{-5} e^{8.933 \times Stretch} M_{\odot}$ and $M_{Ni,3} = 4.934 \times 10^{-7} e^{6.843 \times Stretch} M_{\odot}$, respectively

2.6.2 Time Evolution of Elemental Abundances in SN2011fe and SN2013dy

Among all six SNe with spectroscopic observations near B-maximum and wavelength coverage over 2000–10,000 Å, there are three spectra of SN 2011fe and four spectra of SN 2013dy that are compatible with our neural network setups. In this section, we investigate the time evolution of the elemental abundances of this two SNe. The time evolution of the masses of radioactive materials can serve as an important check on the fidelity of the elemental abundances we deduce from the neural network whose performance is very difficult to track precisely from first principle mathematical models.

From Fig. 2.18a–c, we notice a general agreement between the Ni masses in Zones 2, 3, and 4, respectively, and the radioactive decay of ^{56}Ni. This agreement suggests that most of the ^{56}Ni are newly synthesized after the explosion. SN 2013dy shows a significantly larger Nickel mass than SN 2011fe.

Figure 2.18d–f shows the masses of Co in Zones 2, 3, and 4, respectively. Again, SN 2013by is more abundant in Co in Zones 3 and 4 than SN 2011fe; the abundance ratio seems to be consistent with $\sim$1 in Zone 1. Notice that the mass of Co peaks at around 20 days past explosion if its time variation is related to the radioactive decays of ^{56}Ni; Fig. 2.18d, e is in agreement with this. Likewise, more iron is found

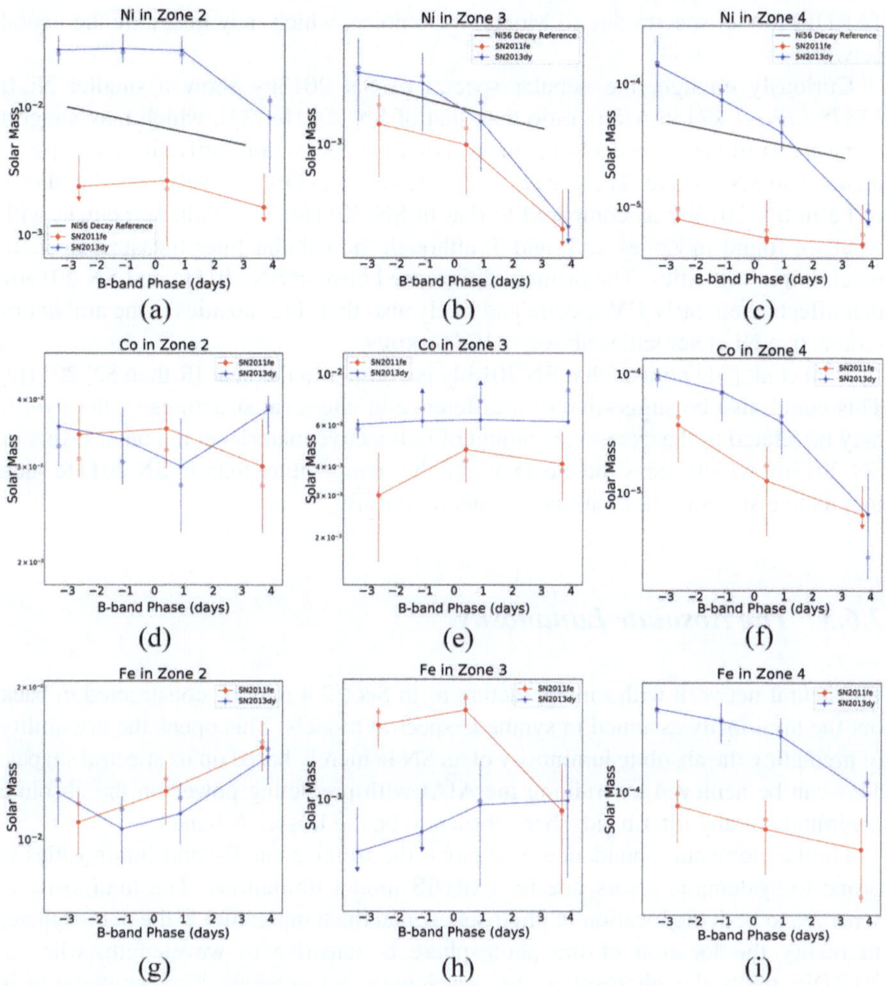

Fig. 2.18 The time evolution of Ni mass in different zones of SN2011fe and SN2013dy near B-band maximum dates. **Upper panel**: From left to right: Ni mass in Zones 2 (**a**), 3 (**b**), and 4 (**c**). The decay rate of Ni_{56} is plotted in black lines for comparison. **Middle panel**: From left to right: Co mass in Zones 2 (**d**), 3 (**e**), and 4 (**f**). **Lower panel**: From left to right: Fe mass in Zones 2 (**g**), 3 (**h**), and 4 (**i**)

at the highest velocity (Zone 3) for SN 2013fe than SN 2011fe. Iron seems to be more abundant in SN 2013by than in SN 2011fe for Zone 2.

Moreover, the predicted Ni masses from the spectra 3–4 days past maximum are significantly lower than earlier epochs. We surmise such anomaly could due to temperature change which strongly affects the UV spectral features. When Fe-group element is in low-temperature region, like Zones 3 or 4, the relevant atomic levels are less activated, which result in weaker or the absence of spectral lines in the

TARDIS model spectra due to Monte Carlo noise, which may misguide the neural network.

Curiously enough, the nebular spectra of SN 2013by show a smaller Ni II λ7378 Å/Fe II λ7155 Å flux ratio than that of SN 2011fe [32], which may suggest that SN 2013dy produced a lower mass ratio of stable to radioactive iron-group elements than SN 2011fe. The decay of ^{56}Ni leads to a comparatively overabundance of Fe in SN 2013by as compared to that in SN 2011fe. This is in agreement with what we found in Zones 2, 3, and 4, although the nebular lines measure ejecta at much lower velocities. The primary difference between SN 2011fe and SN 2013by that affects their early UV spectra and likely also their luminosities is the amount of radioactive Ni at velocities above $\sim$10,000 km/s.

Zhai et al. [31] noticed that SN 2013dy is dimmer in the near IR than SN 2011fe. This could also be suggestive of a difference in chemical structures of the two. It may be related to the excessive amount of radioactive material at the outer layers of SN 2013dy which keeps the ejecta at a higher temperature than in SN 2011fe such that its fine structure lines are not excited in the IR.

2.6.3 The Absolute Luminosity

The neural network with same structure as in Sect. 2.4 may be constructed to back out the luminosity assumed in synthetic spectral models. This opens the possibility of predicting the absolute luminosity of an SN Ia merely based on its spectral shapes. This can be achieved by training the AIAI with predicting power on the absolute magnitude of any filter bands, here chosen to be the Bessel B-band.

Unlike elemental abundance predictors, the training on B-band luminosities is prone to systematic errors due to TARDIS model limitations. The luminosity is sensitive to both the location of photosphere and the temperature at the photosphere. In reality the location of the photosphere is sensitive to wavelength, whereas TARDIS treats the photosphere as wavelength independent. The temperature is sensitive to the UV fluxes, and the location of the photosphere is sensitive to the absorption minimum of weak lines. At around optical maximum, the photosphere has receded into the iron-rich layers for normal SN Ia, and the absorption minima of intermediate mass elements are no longer good indicators of the location of the photosphere. Such insensitivity may introduce degeneracy in the dependence of luminosity and spectral profiles: The luminosity of two supernovae may be very different, whereas the spectral profiles in the optical may appear to be very similar.

Nonetheless, one may try to study the relations between the luminosities and spectral shapes, in a similar way to what have been done for elemental abundances. However, when the neural networks are trained with multiple iterations, the MSE on the training set decreases, while the MSE on the testing set increases. This indicates

Table 2.7 B-band absolute luminosity comparisons. **Column names:** SN name: The name of supernovae. Phase: The days relative to B-band maximum time. Obs. M: The observed magnitude, interpolated from the photometry using SNooPy. $E(B-V)$: The total extinction parameters, including both Milky-Way and host galaxy extinction. μ: The distance modulus, data sources are listed in Table 2.4. MRNN M: The absolute magnitude predicted by our MRNN, we adopt the median values and the $1-\sigma$ intervals in testing dataset. Abs. M: The absolute magnitude calculated from the observational magnitude. Δ M: The difference between MRNN predicted absolute magnitude and the observational absolute magnitude

SN name	Phase	Obs. M	$E(B-V)$	μ	MRNN M	Abs. M	Δ M
SN2011fe	−2.6	10.034	0	29.13	$-19.96^{+0.12}_{-0.12}$	−19.10	0.86
SN2011fe	0.4	9.976	0	29.13	$-20.00^{+0.11}_{-0.11}$	−19.15	0.85
SN2011fe	3.7	10.070	0	29.13	$-19.96^{+0.11}_{-0.12}$	−18.96	0.90
SN2013dy	−3.1	13.353	0.341	31.50	$-20.06^{+0.08}_{-0.09}$	−19.54	0.52
SN2013dy	−1.1	13.294	0.341	31.50	$-20.08^{+0.10}_{-0.09}$	−19.60	0.48
SN2013dy	0.9	13.291	0.341	31.50	$-20.02^{+0.10}_{-0.09}$	−19.61	0.41
SN2013dy	3.9	13.366	0.341	31.50	$-19.99^{+0.11}_{-0.11}$	−19.53	0.46
SN2011by	−0.4	12.933	0.052	31.59	$-20.03^{+0.09}_{-0.09}$	−18.87	1.16
SN2011iv	0.6	12.484	0	31.26	$-19.87^{+0.10}_{-0.09}$	−18.82	1.05
SN2015F	−2.3	13.590	0.210	31.89	$-19.89^{+0.11}_{-0.10}$	−19.16	0.73
ASASSN-14lp	−4.4	12.496	0.351	30.84	$-20.04^{+0.09}_{-0.09}$	−19.74	0.30

that the neural network overfits features in the training set, and the results cannot be used for model predictions. Consequently, we curtailed the performance tolerance to 1 iteration only and adopted a smaller learning rate, which is one-tenth of the value used for training the elemental abundances (3×10^{-6} in the first stage and 3×10^{-8} in the second stage).

Having done the training and testing, we insert the observed spectra into the trained neural network. The luminosities of 11 spectra of 6 SNe were predicted by the neural network and thus constructed; the results are listed in Table 2.7.

We have also estimated the B-band maximum luminosities of the 6 SNe using their light curves with SNooPy [53]. They are converted into absolute magnitude as shown in Table 2.7.

The B-band magnitudes from the neural network are systematically more luminous than the values deduced from well-calibrated light curves. This difference is perhaps the result of the simplified assumption of a sharp photosphere as mentioned above. Despite this limitation, the neural network prediction is largely a measurement of the spectral properties and can still be used as an empirical indicator of luminosity, which is still useful in exploring the diversity of SN Ia based on spectroscopic data (Table 2.8).

Table 2.8 *B*-band absolute luminosity comparisons of 15 SNe from [28]. Due to the lack of photometric data of these SNe, we did not present the possible luminosities for the relating spectra as comparison

Supernova name	Phase	Stretch	MRNN luminosity
PTF09dlc	2.8	1.05 ± 0.03	$-19.80^{+0.12}_{-0.08}$
PTF09dnl	1.3	1.05 ± 0.02	$-19.91^{+0.11}_{-0.1}$
PTF09fox	2.6	0.92 ± 0.04	$-19.94^{+0.11}_{-0.11}$
PTF09foz	2.8	0.87 ± 0.06	$-19.78^{+0.13}_{-0.08}$
PTF10bjs	1.9	1.08 ± 0.02	$-19.89^{+0.11}_{-0.1}$
PTF10hdv	3.3	1.05 ± 0.07	$-19.85^{+0.1}_{-0.09}$
PTF10hmv	2.5	1.09 ± 0.01	$-19.80^{+0.12}_{-0.08}$
PTF10icb	0.8	0.99 ± 0.03	$-19.93^{+0.12}_{-0.11}$
PTF10mwb	−0.4	0.94 ± 0.03	$-19.87^{+0.11}_{-0.09}$
PTF10pdf	2.2	1.23 ± 0.03	$-19.73^{+0.15}_{-0.08}$
PTF10qjq	3.5	0.96 ± 0.02	$-19.92^{+0.11}_{-0.1}$
PTF10tce	3.5	1.07 ± 0.02	$-19.67^{+0.13}_{-0.11}$
PTF10ufj	2.7	0.95 ± 0.02	$-19.75^{+0.14}_{-0.08}$
PTF10xyt	3.2	1.07 ± 0.04	$-19.83^{+0.11}_{-0.09}$
SN2009le	0.3	1.08 ± 0.01	$-19.84^{+0.1}_{-0.09}$

References

1. Parrent J, Branch D, Jeffery D (Oct. 2010) SYNOW: a highly parameterized spectrum synthesis code for direct analysis of SN spectra
2. Thomas RC, Nugent PE, Meza JC (Feb. 2011) SYNAPPS: data-driven analysis for supernova spectroscopy. Publ Astron Soc Pac 123:237
3. Kerzendorf WE, Sim SA (2014) A spectral synthesis code for rapid modelling of supernovae. Mon Not R Astron Soc 440:387–404
4. Lucy LB (May 1999) Improved Monte Carlo techniques for the spectral synthesis of supernovae. Astron Astrophys 345:211–220
5. Lucy LB (Mar 2002) Monte Carlo transition probabilities. Astron Astrophys 384:725–735
6. Khokhlov AM (May 1991) Delayed detonation model for type IA supernovae. Astron Astrophys 245:114–128
7. Hoeflich P, KhokhlovA, Wheeler JC, Phillips MM, Suntzeff NB, Hamuy M (Dec 1996) Maximum brightness and postmaximum decline of light curves of type IA supernovae: a comparison of theory and observations. Astrophys J Lett 472:L81
8. Nomoto K, Thielemann FK, Yokoi K (Nov 1984) Accreting white dwarf models of Type I supernovae. III - carbon deflagration supernovae. Astrophys J 286:644–658
9. Mazzali PA, Sullivan M, Hachinger S, Ellis RS, Nugent PE, Howell DA, Gal-Yam A, Maguire K, Cooke J, Thomas R, Nomoto K, Walker ES (Apr 2014) Hubble space telescope spectra of the type Ia supernova SN 2011fe: a tail of low-density, high-velocity material with Z $Z_\odot$. Mon Not R Astron Soc 439:1959–1979
10. Garavini G, Nobili S, Taubenberger S, Pastorello A, Elias-Rosa N, Stanishev V, Blanc G, Benetti S, Goobar A, Mazzali PA, Sanchez SF, Salvo M, Schmidt BP, Hillebrandt W (Aug 2007) ESC observations of SN 2005cf. II. Optical spectroscopy and the high-velocity features. Astron Astrophys 471:527–535
11. Hillier DJ, Miller DL (Mar 1998) The treatment of non-LTE line blanketing in spherically expanding outflows. Astrophys J 496:407–427
12. Abdi M, Nahavandi S (2016) Multi-residual networks. CoRR, abs/1609.05672
13. Savitzky A, Golay MJE (1964) Smoothing and differentiation of data by simplified least squares procedures. Anal Chem 36(8), 1627–1639

14. LeCun Y , Bengio Y, Hinton G (2015) Deep learning. Nature 521:436–444
15. Montúfar G, Pascanu R, Cho K, BengioY (Feb 2014) On the number of linear regions of deep neural networks. arXiv e-prints, arXiv:1402.1869
16. Krizhevsky A , Sutskever I, Hinton GE (2012) ImageNet classification with deep convolutional neural networks. In: Pereira, F, Burges CJC, Bottou L, Weinberger KQ (eds) Advances in neural information processing systems 25. Curran Associates, Inc., Red Hook, pp 1097–1105
17. Simonyan K, Zisserman A (Sep 2014) Very deep convolutional networks for large-scale image recognition. arXiv e-prints, arXiv:1409.1556
18. Ioffe S, Szegedy C (Feb 2015) Batch normalization: accelerating deep network training by reducing internal covariate shift. arXiv e-prints, arXiv:1502.03167
19. He K, Zhang X, Ren S, Sun J (Dec 2015) Deep residual learning for image recognition. arXiv e-prints, arXiv:1512.03385
20. Huang G, Liu Z, van der Maaten L, Weinberger KQ (Aug 2016) Densely connected convolutional networks. arXiv e-prints, arXiv:1608.06993
21. Krizhevsky A (2012) Learning multiple layers of features from tiny images. University of Toronto
22. Chollet F et al (2015) Keras. https://keras.io
23. Abadi M , Barham P, Chen J, Chen Z, Davis A, Dean J , Devin M, Ghemawat S, Irving G, Isard M , et al (2016) TensorFlow: a system for large-scale machine learning. In: 12th {USENIX} symposium on operating systems design and implementation ({OSDI} 16), pp 265–283
24. Kingma DP, Ba J (2014) Adam: a method for stochastic optimization. CoRR, abs/1412.6980
25. Foreman-Mackey D, Hogg DW, Lang D, Goodman J (Mar 2013) emcee: the MCMC hammer. Publ Astron Soc Pacific 125:306
26. Graur O, Zurek DR, Rest A, Seitenzahl IR, Shappee BJ, Fisher R, Guillochon J, Shara MM, Riess AG (June 2018) Observations of SN 2015F suggest a correlation between the intrinsic luminosity of type Ia supernovae and the shape of their light curves ~900 days after explosion. Astrophys J 859:79
27. Munari U, Henden A, Belligoli R, Castellani F, Cherini G, Righetti GL, Vagnozzi A (Apr 2013) BVRI lightcurves of supernovae SN 2011fe in M101, SN 2012aw in M95, and SN 2012cg in NGC 4424. New Astron 20:30–37
28. Maguire K, Sullivan M, Ellis RS, Nugent PE, Howell DA, Gal-Yam A, Cooke J, Mazzali P, Pan YC, Dilday B, Thomas RC, Arcavi I, Ben-Ami S, Bersier D, Bianco FB, Fulton BJ, Hook I, Horesh A, Hsiao E, James PA, Podsiadlowski P, Walker ES, Yaron O, Kasliwal MM, Laher RR, Law NM, Ofek EO, Poznanski D, Surace J (Nov 2012) Hubble Space Telescope studies of low-redshift Type Ia supernovae: evolution with redshift and ultraviolet spectral trends. Mon Not R Astron Soc 426:2359–2379
29. Graham ML, Foley RJ, Zheng W, Kelly PL, Shivvers I, Silverman JM, Filippenko AV, Clubb KI, Ganeshalingam M (Jan 2015) Twins for life? A comparative analysis of the Type Ia supernovae 2011fe and 2011by. Mon Not R Astron Soc 446:2073–2088
30. Foley RJ, Hoffmann SL, Macri LM, Riess AG, Brown PJ, Filippenko AV, Graham ML, Milne PA (June 2018) Significant luminosity differences of two twin type Ia supernovae. arXiv e-prints
31. Zhai Q, Zhang JJ, Wang XF, Zhang TM, Liu ZW, Brown PJ, Huang F, Zhao XL, Chang L, Yi WM, Wang CJ, Xin YX, Wang JG, Lun BL, Zhang XL, Fan YF, ZhengXM, Bai JM (May2016) UV-optical observation of type Ia supernova SN 2013dy in NGC 7250. Astron J 151:125
32. Pan YC, Foley RJ, Kromer M, Fox OD, Zheng W , Challis P, Clubb KI, Filippenko AV, Folatelli G, Graham ML, Hillebrandt W, Kirshner RP, Lee WH, Pakmor R, Patat F, Phillips MM, Pignata G, Röpke F, Seitenzahl I, Silverman JM, Simon JD, Sternberg A, Stritzinger MD, Taubenberger S, Vinko J, Wheeler JC (2015) 500 days of SN 2013dy: spectra and photometry from the ultraviolet to the infrared. Mon Not R Astron Soc 452:4307–4325

33. Foley RJ, Kromer M, Howie Marion G, Pignata G, Stritzinger MD, Taubenberger S, Challis P, Filippenko AV, Folatelli G, Hillebrandt W, Hsiao EY, Kirshner RP, Li W, Morrell NI, Röpke FK, Ciaraldi-Schoolmann F, Seitenzahl IR, Silverman JM, Simcoe RA, Berta ZK, Ivarsen KM, Newton ER, Nysewander MC, Reichart DE (July 2012) The first maximum-light ultraviolet through near-infrared spectrum of a type Ia supernova. Astrophys J Lett 753:L5
34. Shappee BJ, Piro AL, Holoien TWS, Prieto JL, Contreras C, Itagaki K, Burns CR, Kochanek CS, Stanek KZ, Alper E, Basu U, Beacom JF, Bersier D, Brimacombe J, Conseil E, Danilet AB, Dong S, Falco E, Grupe D, Hsiao EY, Kiyota S, Morrell N, Nicolas J, Phillips MM, Pojmanski G, Simonian G, Stritzinger M, Szczygieł DM, Taddia F, Thompson TA, Thorstensen J, Wagner MR, Woźniak PR (Aug 2016) The young and bright type Ia supernova ASASSN-14lp: discovery, early-time observations, first-light time, distance to NGC 4666, and progenitor constraints. Astrophys J 826:144
35. Graur O, Zurek DR, Cara M, Rest A, Seitenzahl IR, ShappeeBJ, Shara MM, Riess AG (Oct 2018) Late-time observations of ASASSN-14lp strengthen the case for a correlation between the peak luminosity of type Ia supernovae and the shape of their late-time light curves. Astrophys J 866:10
36. Nugent PE, Sullivan M, Cenko SB, Thomas RC, Kasen D, Howell DA, Bersier D, Bloom JS, Kulkarni SR, Kandrashoff MT, Filippenko AV, Silverman JM, Marcy GW, Howard AW, Isaacson HT, Maguire K, Suzuki N, Tarlton JE, Pan YC, Bildsten L, Fulton BJ, Parrent JT, Sand D, Podsiadlowski P, Bianco FB, Dilday B, Graham ML, Lyman J, James P, Kasliwal MM, Law NM, Quimby RM, Hook IM, Walker ES, Mazzali P, Pian E, Ofek EO, Gal-Yam A, Poznanski D (Dec 2011) Supernova SN 2011fe from an exploding carbon-oxygen white dwarf star. Nature 480:344–347
37. Chomiuk L, Soderberg AM, Moe M, Chevalier RA, Rupen MP, Badenes C, Margutti R, Fransson C, Fong W-f, Dittmann JA (May 2012) EVLA observations constrain the environment and progenitor system of type Ia supernova 2011fe. Astrophys J 750:164
38. Patat F, Cordiner MA, Cox NLJ, Anderson RI, Harutyunyan A, Kotak R, Palaversa L, Stanishev V, Tomasella L, Benetti S, Goobar A, Pastorello A, Sollerman J (Jan. 2013) Multi-epoch high-resolution spectroscopy of SN 2011fe. Linking the progenitor to its environment. Astron Astrophys 549:A62
39. Gall C, Stritzinger MD, Ashall C, Baron E, Burns CR, Hoeflich P, Hsiao EY, Mazzali PA, Phillips MM, Filippenko AV, Anderson JP, Benetti S, Brown PJ, Campillay A, Challis P, Contreras C, Elias de la Rosa N, Folatelli G, Foley RJ, Fraser M, Holmbo S, Marion GH, Morrell N, Pan Y-C, Pignata G, Suntzeff NB, Taddia F, Torres Robledo S, Valenti S (Mar. 2018) Two transitional type Ia supernovae located in the Fornax cluster member NGC 1404: SN 2007on and SN 2011iv. Astron Astrophys 611:A58
40. Filippenko AV, Silverman JM, Ganeshalingam M (2013) Berkeley supernova Ia program – V. Late-time spectra of type Ia supernovae. Mon Not R Astron Soc 430:1030–1041
41. Fitzpatrick EL (Jan. 1999) Correcting for the effects of interstellar extinction. Publ Astron Soc Pacif 111:63–75
42. Gordon KD, Cartledge S, Clayton GC (Nov. 2009) FUSE measurements of far-ultraviolet extinction. III. The dependence on R(V) and discrete feature limits from 75 galactic sightlines. Astrophys J 705:1320–1335
43. Cartier R, Sullivan M, Firth RE, Pignata G, Mazzali P, Maguire K, Childress MJ, Arcavi I, Ashall C, Bassett B, Crawford SM, Frohmaier C, Galbany L, Gal-Yam A, Hosseinzadeh G, Howell DA, Inserra C, Johansson J, Kasai EK, McCully C, Prajs S, Prentice S, Schulze S, Smartt SJ, Smith KW, Smith M, Valenti S, Young DR (Feb. 2017) Early observations of the nearby Type Ia supernova SN 2015F. Mon Not R Astron Soc 464:4476–4494
44. Cardelli JA, Clayton GC, Mathis JS (Oct. 1989) The relationship between infrared, optical, and ultraviolet extinction. Astrophys J 345:245–256
45. Vinkó J, Ordasi A, Szalai T, Sárneczky K, Bányai E, Bíró IB, Borkovits T, Hegedüs T, Hodosán G, Kelemen J, Klagyivik P, Kriskovics L, Kun E, Marion GH, Marschalkó G, Molnár L, Nagy AP, Pál A, Silverman JM, Szakáts R, Szegedi-Elek E, Székely P, Szing A, Vida K, Wheeler JC (June 2018) Absolute distances to nearby type Ia supernovae via light curve fitting methods. Publ Astron Soc Pacif 130:064101

46. Conley A, Sullivan M, Hsiao EY, Guy J, Astier P, Balam D, Balland C, Basa S, Carlberg RG, Fouchez D, Hardin D, Howell DA, Hook IM, Pain R, Perrett K, Pritchet CJ, Regnault N (Jul 2008) SiFTO: an empirical method for fitting SN Ia light curves. Astrophys J 681:482–498
47. Perlmutter S, Aldering G, Goldhaber G, Knop R, Nugent P, Castro P, Deustua S, Fabbro S, Goobar A, Groom D, et al (1999) Measurements of ω and λ from 42 high-redshift supernovae. Astrophys J 517(2):565
48. Phillips MM (Aug. 1993) The absolute magnitudes of Type IA supernovae. Astrophys J Lett 413:L105–L108
49. Rau A, Kulkarni SR, Law NM, Bloom JS, Ciardi D, Djorgovski GS, Fox DB, Gal-Yam A, Grillmair CC, Kasliwal MM, Nugent PE, Ofek EO, Quimby RM, Reach WT, Shara M, Bildsten L, Cenko SB, Drake AJ, Filippenko AV, Helfand DJ, Helou G, Howell DA, Poznanski D, Sullivan M (Dec. 2009) Exploring the optical transient sky with the palomar transient factory. Publ Astron Soc Pacif 121:1334
50. Law NM, Kulkarni SR, Dekany RG, Ofek EO, Quimby RM, Nugent PE, Surace J, Grillmair CC, Bloom JS, Kasliwal MM, Bildsten L, Brown T, Cenko SB, Ciardi D, Croner E, Djorgovski SG, van Eyken J, Filippenko AV, Fox DB, Gal-Yam A, Hale D, Hamam N, Helou G, Henning J, Howell DA, Jacobsen J, Laher R, Mattingly S, McKenna D, Pickles A, Poznanski D, Rahmer G, Rau A, Rosing W, Shara M, Smith R, Starr D, Sullivan M, Velur V, Walters R, Zolkower J (Dec. 2009) The palomar transient factory: system overview, performance, and first results. Publ Astron Soc Pacif 121:1395
51. Hoeflich P, Hsiao EY, Ashall C, Burns CR, Diamond TR, Phillips MM, Sand D, Stritzinger MD, Suntzeff N, Contreras C, Krisciunas K, Morrell N, Wang L (Sep 2017) Light and color curve properties of type Ia supernovae: theory versus observations. Astrophys J 846:58
52. Shen KJ, Boubert D, Gänsicke BT, Jha SW, Andrews JE, Chomiuk L, Foley RJ, Fraser M, Gromadzki M, Guillochon J, Kotze MM, Maguire K, Siebert MR, Smith N, Strader J, Badenes C, Kerzendorf WE, Koester D, Kromer M, Miles B, Pakmor R, Schwab J, Toloza O, Toonen S, Townsley DM, Williams BJ (Sept. 2018) Three hypervelocity white dwarfs in Gaia DR2: evidence for dynamically driven double-degenerate double-detonation type Ia supernovae. Astrophys J 865:15
53. Burns CR, Stritzinger M, Phillips MM, Kattner S, Persson SE, Madore BF, Freedman WL, Boldt L, Campillay A, Contreras C, Folatelli G, Gonzalez S, Krzeminski W, Morrell N, Salgado F, Suntzeff NB (Dec. 2010) The Carnegie Supernova Project: light-curve fitting with SNooPy. Astron J 141:19

Chapter 3
Artificial Intelligence Assisted Inversion (AIAI): Quantifying the Spectral Features of ^{56}Ni of Type Ia Supernovae

Abstract Following the previous study of artificial intelligence assisted inversion (AIAI) of supernova analyses, we train a set of deep neural networks based on the one-dimensional radiative transfer code TARDIS to simulate the optical spectra of type Ia supernovae (SNe Ia) between 10 and 40 days after the explosion. The neural networks are applied to derive the mass of ^{56}Ni in velocity ranges above the photosphere for a sample of 124 well-observed SNe Ia in the TARDIS model context. A subset of the SNe has multi-epoch observations for which the decay of the radioactive ^{56}Ni can be used to test the AIAI quantitatively. The ^{56}Ni mass derived from AIAI using the observed spectra as input for this subset agrees with the radioactive decay rate of ^{56}Ni. The AIAI reveals that a spectral signature near 3890 Å is related to the Ni II 4067Å line, and the ^{56}Ni mass deduced from AIAI is found to be correlated with the light curve shapes of SNe Ia, with the SNe Ia with broader light curves showing larger ^{56}Ni mass in the envelope above the photosphere. AIAI enables spectral data of SNe to be quantitatively analyzed under theoretical frameworks based on well-defined physical assumptions.

3.1 Methodology

3.1.1 TARDIS Spectral Calculation

In TARDIS, the SN ejecta is assumed to be spherically symmetric and expands homogeneously after the explosion. To avoid the computationally demanding time-dependent photon diffusion calculation, a simple approximation of the photosphere with a black-body emission spectrum at the inner boundary of the SN ejecta is

---

Reprinted with permission from "Artificial Intelligence Assisted Inversion (AIAI): Quantifying the Spectral Features of ^{56}Ni of Type Ia Supernovae" by Xingzhuo Chen, Lei Hu, Lifan Wang, Peter J Brown. 2024. The Astrophysical Journal, Vol. 962, Issue 2, Number 125. https://doi.org/10.3847/1538-4357/ad0a33.

X. Chen, *Deep Learning on Type Ia Supernovae*, Springer Theses,
https://doi.org/10.1007/978-3-032-13032-7_3

adopted. Given an SN ejecta structure, a target luminosity, a time after the explosion, and an inner boundary velocity, TARDIS can calculate a spectrum in $\sim$1 CPU hour.

The model spectra are calculated using the code TARDIS (version 3.0-dev3448). To set up the models, we use dilute local thermodynamic equilibrium (`dilute-lte`) [1] approximation to calculate the atomic level population, nebular approximation (`nebular`) [2] to calculate the ionization fraction, `string macroatom` [3, 4] to calculate the atomic level transition, and `detailed` to calculate the black-body radiative rates. TARDIS uses an iterative algorithm [4] to calculate a self-consistent SN temperature profile. In each iteration, the plasma temperature is modified to satisfy the local energy conservation of the absorbed and emitted photons, and the inner boundary temperature is modified to satisfy the target luminosity. As in our previous work on AIAI of SN Ia spectra (XLL20), the step of temperature iterations is set to 15, with the damping parameters being (`type:damped. damping_constant:1 threshold:0.05. fraction:0.8. hold_iterations:3. t_inner_damping_constant:1`). The temperature calculation setup in our research may not return a converged temperature profile; the output spectrum may not reach the target luminosity. According to XLL20, the B-band absolute luminosity for an SN 2011fe-like model could be $\sim$0.8 Mag brighter than the observation. However, the final temperature profile is dependent on the target luminosity and the location of the inner boundary with the current iterative convergence strategy. We adopt this configuration for its flexibility in the temperature calculation, necessary when the observed and the simulated spectra are compared quantitatively.

3.1.2 Simulated Spectral Library

In this section, we parameterize the supernova ejecta models based on the "IGM" model in XLL20 and randomly sample the parameter space to create a supernova ejecta structure library.

We treat the SN ejecta density profile as a power-law relation shown below:

$$\begin{aligned}\rho(v) = 4.712 \times 10^{-14} &\left(\frac{t}{11.5741\,\text{days}}\right)^{-3} A \\ &\times 0.689^{\frac{(v-12500\,\text{km/s})B}{1000\,\text{km/s}}}\ \text{g/cm}^3,\end{aligned} \tag{3.1}$$

where A and B are dimensionless parameters for the density profile, v is the expansion velocity, and t is the time after the explosion. When $A = 1$ and $B = 1$, the density profile is similar to the "IGM" model in XLL20 and serves as the reference density profile in this chapter. A comparison between the density profiles is shown in Fig. 3.1. We notice the power-law density profile could approximately fit most

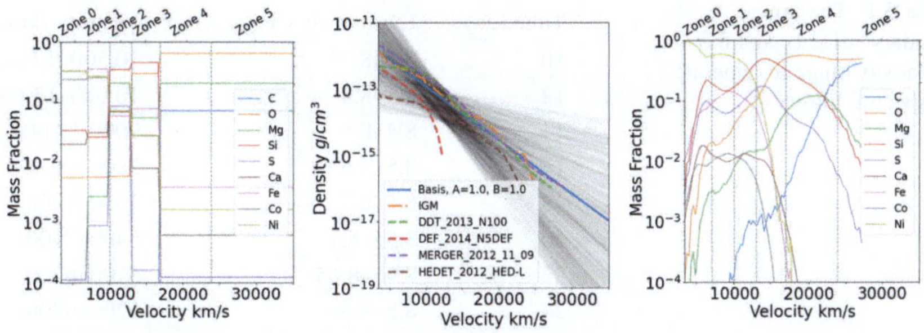

Fig. 3.1 Left: The elemental abundances of the reference model. The labels of the zones are shown on the top of the panel, and the zone boundaries are marked with gray-dashed lines. Middle: The density structure of the reference model with $A = 1$ and $B = 1$ in Eq. 3.1 used in this chapter (blue line), the density structure in XLL20 [5] (orange dash-dotted line). Several other density profiles are also shown in dashed lines with different colors; they are DDT_2013_N100 [6] (green), DEF_2014_N5DEF [7] (red), MERGER_2012_11_09 [8] (purple), and HEDET_2012_HED-L [9] (brown). The models with parameters A changing from 0.2 to 2 and B changing from 0.2 to 2 are shown in gray color. The densities shown are at 11.5741 days after the explosion. Right panel: The elemental abundance structure of the DDT_2013_N100 model at 100 seconds after the explosion

of the theoretical SN ejecta density profiles, while a few theoretical density profiles (i.e., DEF_2014_N5DEF) require more parameters to be fitted.

We divide the SN ejecta into six discrete zones in the velocity space. The elemental abundance is assumed to be constant inside each zone. There is no layer above 24,000 km/s in XLL20; we extrapolate the high-velocity component from the "IGM" model in XLL20 to serve as the reference elemental abundance in this chapter, as is shown in the left panel of Fig. 3.1. The elemental abundance structure and the density profile of the DDT_2013_N100 model from [6] are also shown in Fig. 3.1 for comparison.

The parameters for generating the spectral library include:

- The density profile parameterized by A and B (see Eq. 3.1): A and B are drawn from uniform distributions between [0.2, 2] and [0, 2], respectively.
- The abundance of 23 elements in six zones of the ejecta structure: The abundance of an element in a given zone is drawn from a Gaussian distribution in the logarithmic space $N(\mu, \sigma^2)$, where μ is the log of the abundance of the reference elemental abundance structure, and σ is set to be 0.5 for all elements. Moreover, the sum of the elemental abundance in each zone is conserved to be 1.
- The time after the explosion: The time after the explosion is drawn from a uniform distribution in [10, 40] days.
- The supernova luminosity between 6500 Å and 7500 Å in units of $log_{10}(L_{\odot})$ and the location of the inner boundary: To set the parameter ranges of the SN luminosity and the location of the photosphere, we first use the reference ejecta structure (with $A = B = 1$ in Eq. 3.1) and the elemental abundance in Fig. 3.1 to calculate a spectral sequence between 10 days and 40 days after the explosion,

Table 3.1 The inner boundary velocity and the luminosity range to generate the spectral library

Time (days)	Luminosity ($log_{10}(L_{\odot})$)	Velocity (km/s)
10	8.2–8.35	10,500–13,000
12	8.3–8.4	10,000–13,000
14	8.4–8.5	9000-12,000
16	8.5–8.6	7800–10,000
18	8.5–8.7	7000–8200
20	8.55–8.7	6400–7500
22	8.55–8.65	5700–6700
24	8.5–8.65	5200–6200
26	8.47–8.62	4500–5800
28	8.42–8.6	4300–5100
30	8.38–8.52	4000–4900
32	8.3–8.5	3750–4700
34	8.15–8.45	3500–4700
36	8.1–8.35	3500–4700
38	8.0–8.3	3500–4700
40	7.8–8.2	3500–4700

with varying target luminosities and inner boundary velocities. By comparing the simulated spectra with the observed SNe Ia spectral energy distribution [10, 11], we determine the lower and upper limits of the target luminosities and the locations of the inner boundary at different epochs after the explosion, and the results are shown in Table 3.1.

When generating spectra for the spectral library to be used for NN training, we sample the day after the SN explosion from a uniform distribution in $[10, 40]$ and then sample a target luminosity in logarithmic space from the uniform distribution between $[L_{min}, L_{max}]$, where L_{min} and L_{max} are taken from Table 3.1 with linear interpolations for days not included in the table. The inner boundary velocity is also sampled from the uniform distribution in $[V_{min}, V_{max}]$, where V_{min} and V_{max} are linearly interpolated from Table 3.1.

The wavelength range of the model spectra is between 2000 and 10,000 Å with a total of 2000 wavelength bins. Each wavelength bin is separated by a frequency difference of 5.9958×10^{11} Hz. A total of 108,389 ejecta models and their corresponding spectra are calculated to form the spectral training and testing sets for building the NNs.

3.1.3 Neural Network

A typical NN consists of an input, an output, and multiple hidden layers. The input layer receives data as input tensors, and the hidden layers process the tensors and send the results to the output layer. Different types of layers, representing different mathematical calculations, can be used in the hidden layers. For example, the convolution layer convolves the data tensor with a convolution core, the activation layer applies a nonlinear function onto the data tensor, the pooling layer reduces the size of the data tensor by binning the adjacent data with the maximum values or the average values, and the fully connected layer multiplies a matrix to the input tensor (for a review, see [12]).

The goal of the neural network is to provide predictions (y_{pred}) over a variable of our interest that is close to the truth (y_{true}) from the given information (X). Initially, the values of the trainable parameters (i.e., the convolution core in the convolution layers and the matrix in the fully connected layers, all denoted as w) are randomly assigned. The neural network then calculates the output (y_{pred}) using the known information X as input, and a loss function (i.e., mean squared error: $Loss = Mean(y_{true} - y_{pred})^2)$ is calculated to evaluate the performance of prediction. The trainable parameters are updated with respect to the gradient $\partial L/\partial w$ and a predefined learning rate to minimize the loss function. The training dataset is separated into batches for computational efficiency, and the trainable parameters are updated at each batch. The neural network browses the dataset multiple times; one browse is defined as one epoch.

We adopt the deep ensemble NNs [13], which can make predictions and uncertainties simultaneously. Compared to typical NNs such as AlexNet [14] and VGG16 [15]), the output of deep ensemble NNs is not just the prediction value, but the prediction mean μ together with the prediction error σ, assuming a Gaussian noise distribution. The loss function is

$$Loss = Mean\left(ln(\sigma(x)) + \frac{(y_{true} - \mu(x))^2}{\sigma(x)^2}\right), \tag{3.2}$$

where x is the input information of the neural network, including the spectrum, the time after the explosion, and the density parameters; μ and σ are written in function forms to represent the neural network.

We use the multi-residual NNs [16] for the NN architecture. In brief, we introduce a block structure with two convolution layers, a batch normalization layer and an activation layer, and repeat the block structure seven times. The input of the block structure is the sum of the output of all the previous block structures and the input of the first block structure. Moreover, the NN has two inputs, the 2000-element array `input_1` that receives the input spectra with 2000 pixels from 2000 to 10,000 Å and the 3-element array `input_2` that receives the density parameters A and B shown in Eq. 3.1 and the time after the explosion. We use the scaled exponential linear unit (SELU) function as the activation function for all the hidden layers:

$$selu(x) = \begin{cases} scale \times x, & (x > 0) \\ scale \times \alpha \times (e^x - 1), & (x \leq 0), \end{cases} \tag{3.3}$$

where $scale = 1.05070098$ and $\alpha = 1.67326324$. The detailed architecture is shown in a figure in our GitHub repository at https://geronimochen.github.io/images/Model_1_0.h5.pdf.

Among all 108,389 spectra, 80% are assigned to the training set, and the remaining 20% are in the testing set. The training dataset updates the trainable parameters, while the testing dataset only monitors the loss function after every epoch.

Note that not all the observed spectral data cover the entire UV-optical-IR wavelength from 2000 to 10,000 Å. Data augmentation is applied to the dataset to cope with the problem. In this augmentation method, we duplicate the training and testing datasets by making five copies of them. The spectra in the training and testing datasets' first copies remain the same as those in the original spectral library. For each spectrum in the other four copies, we randomly select an observational facility (telescope and instrument) that was used to acquire the spectral observations of SNe in the WISeREP database and then apply the wavelength coverage of that facility onto the spectrum. Furthermore, each spectrum is normalized by dividing the spectrum by the average pixel value to avoid overflow in the NNs.

We used the `adam` algorithm [17] to update the trainable parameters with a batch size 200. The training process consists of two phases with different learning rates. In the first phase, the learning rate is 0.00003 and decays 10^{-6} per batch. The loss in Eq. 3.2 is calculated on the training and testing datasets, respectively, after every epoch. When the loss on the testing dataset does not decrease in the recent 10 epochs, we cease the first phase and keep the model with the smallest loss for the second phase. In the second phase, the learning rate is 0.0000003 and decays 10^{-6} per batch. We also monitor the losses on the training and testing datasets at every epoch. When the loss on the testing dataset does not decrease for 10 consecutive epochs, we cease the training and keep the NN model with the smallest loss on the testing dataset.

We trained seven sets of NNs in total; the first six NNs are designed to predict 30 elemental abundances (from H to Zn) in Zones 0–5, and the seventh NN is designed to predict the inner boundary velocity and the luminosity. The seventh NN outputs a two-element vector, while the other six NNs output 30-element vectors. The losses on the testing dataset are shown in Figs. 3.2 and 3.3.

3.1.4 Density Structure

We propose two density estimation methods in this research. In the first method, we assume the density profile of an SN Ia follows Eq. 3.1 with A and B parameters independent of velocity and time after the explosions, then we design a grid-search method to find the simulated spectral time sequence that best matches the observed spectral time sequence in the parameter space $0.2 \leq A \leq 2$ and $0 \leq B \leq 2$, and the spacing is 0.1 for A and B, as is introduced in Sect. 3.3. This method assumes that Eq. 3.1 is universally correct and is apparently an oversimplification. Indeed,

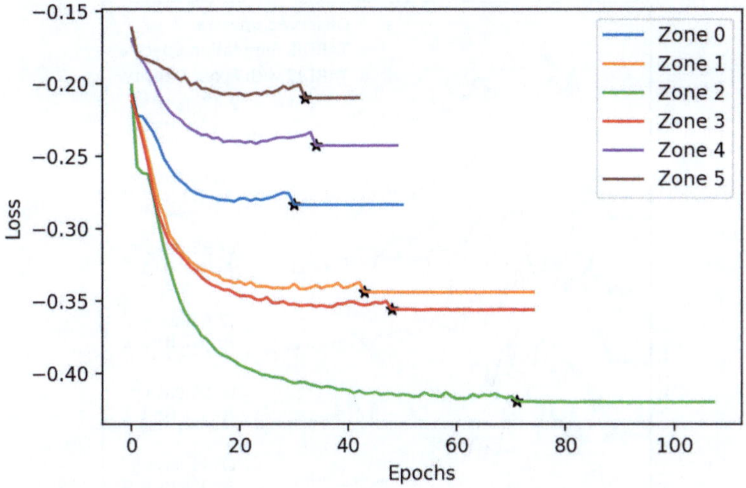

Fig. 3.2 The learning curves of the six NNs for elemental abundance predictions in Zones 0–5, showing the losses on the testing dataset at different epochs. The black stars show the starting points of the second phase in the training process. The losses for Zones 0 to 5 are shown with blue, orange, green, red, purple, and brown lines, respectively

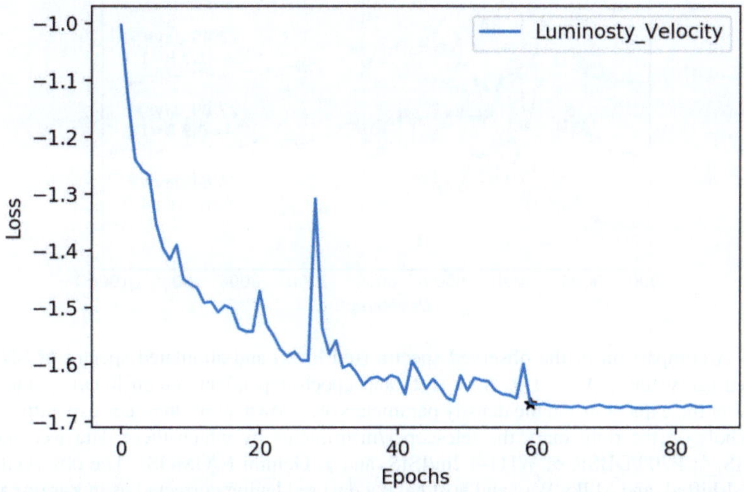

Fig. 3.3 The learning curve of the NN predicting the luminosity and inner boundary velocity. The line shows the losses with the testing dataset at different epochs. The black star shows the beginning of the second phase in the training process

we found the simulated spectra from this method failed to model the entire time sequence of some well-observed SNe Ia, especially in the UV, where the spectra are most sensitive to the density profiles. An example is shown in Fig. 3.4 for SN 2011fe. While the majority of the data are well fitted, the model UV spectra

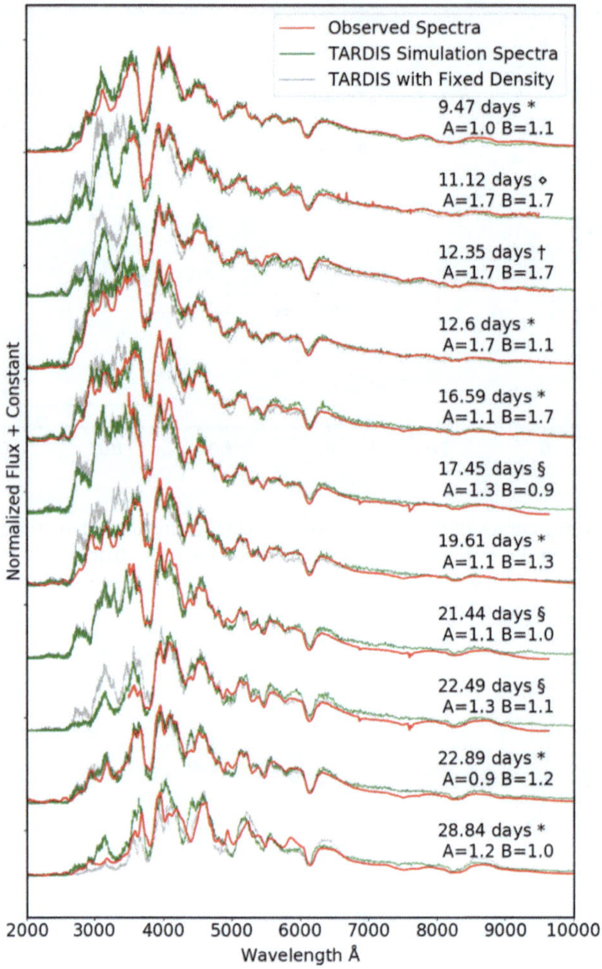

Fig. 3.4 A comparison of the observed spectra (red lines) and simulated spectra of SN 2011fe with fixed (gray lines; $A = 1.0$, $B = 1.1$) and epoch-dependent (green lines) A and B. The epochs after the explosion and the density parameters are shown above their corresponding spectra. The symbols on the right mark the telescopes/instruments by which these data were taken ($*$: HST/STIS, †: P200/DBSP, ⋄: WHT-4.2m/ISIS, and §: Gemini-N/GMOS). The observed spectra are de-redshifted, and Milky Way and host galaxy dust reddening corrected as in `kaepora` [18]

(Fig. 3.4, gray lines) on days 16.59 and 19.61 show significant deviations from the observations (red lines). In reality, at different epochs, different ejecta layers are exposed, and the density profiles may deviate from Eq. 3.1.

We therefore introduce the second method that allows the density parameters A and B to change with time and their values derived using only the spectrum taken at the corresponding epochs; the same A and B grid defined for the first method is used to search for the values of A and B that lead to optimal spectral fits. To assess the goodness of the spectral fit, we use the NNs to predict the chemical structure of the ejecta for each spectrum and each pair of A and B in the grid. A TARDIS model spectrum is calculated using the predicted ejecta structure as input. A mean squared error (MSE) is derived using the following equation to estimate the best match chemical structures and density profile parameters A and B

$$MSE = \sum_{\lambda}(F_{obs}(\lambda) - F_{sim}(\lambda))^2/n \, , \tag{3.4}$$

where $F_{obs}(\lambda)$ and $F_{sim}(\lambda)$ are the observed and simulated spectra normalized by their corresponding means, respectively, and n is the number of wavelength bins. Here, the wavelength grid is resampled according to the simulated spectrum, and the regions with missing observations are set to have zero weights.

The simulated spectra with varying A and B parameters show noticeable improvements in reproducing the observed spectra over the models with fixed A and B, as shown in Fig. 3.4 for SN 2011fe as an example; the fits to days 16.59 and 19.61 are significantly better. However, we note that the difference in density structure is typically small for the two different treatments of A and B and does not lead to significant discrepancies in the abundance estimates, as can be seen in Sects. 3.5 and 3.3.

Overall, the quality of the spectral fits is worse in the ultraviolet than in the visible wavelength. This may be due to several reasons. First, the density profile used in this simulation is a simple exponential function throughout the ejecta, which can be an oversimplification and the primary source of the mismatch in the UV spectral region; the UV flux is extremely sensitive to the ejecta moving at all velocities. Second, the nonlocal thermodynamic equilibrium (NLTE) effect becomes more significant at the late time SNe Ia spectra, while TARDIS only uses the `nebula` approximation to account for this effect. Third, the assumption that all photons initiate from a sharp inner boundary is only approximately correct, which could lead to systematic flux errors, especially in the UV (for a comparative study of TARDIS and other radiative transfer models [19]). Fourth, the high dimensionality of the model parameter space means that the library of theoretical models may not be extensive enough to cover completely all the subtypes of SNe Ia and could be improved with a more comprehensive spectral library. Nonetheless, the AIAI approach, as adopted in this study, makes it possible to derive model-based quantities that are sensitive to the chemical structures of the ejecta, although these derived quantities are dependent on the adopted models and the radiative transfer program.

3.1.5 ^{56}Ni Line Identification

To identify the spectral lines from Ni, we apply our algorithm to the Hubble Space Telescope (HST) spectrum of SN 2011fe at 19.61 days after the explosion to derive the chemical structure of the SN ejecta. Based on this ejecta structure, we

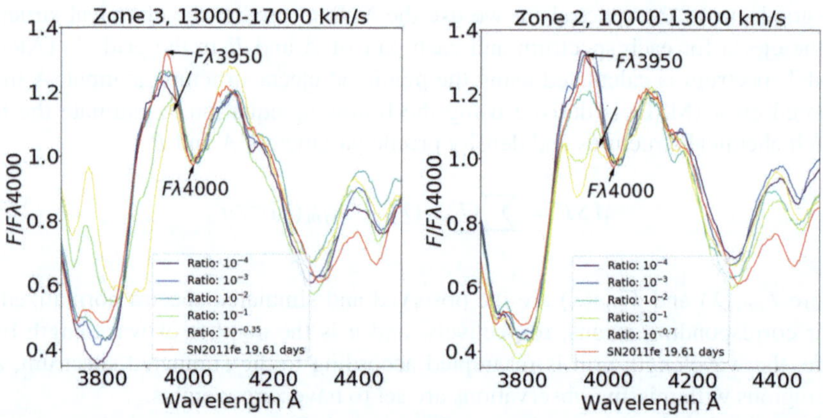

Fig. 3.5 An illustration of measuring the spectral line flux ratio $FRNi$. The red lines are the spectra of SN 2011fe at 19.61 days after the explosion. The lines with different colors are the simulated spectra using TARDIS. The simulated spectra on the left and right are computed with varying Ni mass fractions in the velocity ranges 13,000–17,000 km/s (Zone 3) and 10,000–13,000 km/s (Zone 2), respectively. The Ni mass fractions are shown in the legends of the two figures. All the spectra are normalized to the local minimum flux near 4000 Å. The black arrows indicate the local minimum and maximum that are used to measure the spectral line flux ratio $FRNi = F(\lambda3950)/F(\lambda4000)$

artificially modify the elemental abundance in the velocity range between 13,000 and 17,000 km/s (denoted as Zone 3) and the velocity range 10,000–13,000 km/s (denoted as Zone 2) by changing the Ni fraction, while rescaling the abundances of the rest of the elements to keep the density profile conserved. Figure 3.5 shows the simulated spectra with Ni abundance in Zone 2 and 3 varying from 10^{-4} to $10^{-0.7}$ and 10^{-4} to $10^{-0.35}$, respectively. We notice that Ni strongly affects the spectral features at 3750–3950Å, which can be attributed to the blueshifted absorption of the Ni II 4067 Å transition lines. This is the only line in the optical in which the strength of Ni II absorption can be directly measured. This line is especially prominent when the Ni fraction is above 10^{-2} in these velocity regions. Note also that the pseudo-equivalent width (pEW) of the Si II absorption line at ~4000 Å is affected by the Ni II transition lines. It has been shown in previous studies [20, 21] that the pseudo-equivalent width (pEW) of the Si II absorption line at ~4000 Å is correlated with the light curve stretch parameter and the absolute magnitude of SNe Ia. In reality, the measurement of the pEW of the 4000 Å feature captures the combined effect of the Ni II, Ca II, Fe II, and the Si II lines in this wavelength region. Figure 3.5 demonstrates that the origin of the correlation between the pEW of the 4000 Å feature and the light curve shape and the absolute magnitude of SNe Ia is more likely to be directly associated with the strength of the Ni II 4067 Å line.

We focus on the flux ratio of the emission around 3950 Å and the absorption feature around 4000 Å; the two spectral features can be measured in most of the observed SNe Ia spectra. The wavelength range is small, so the flux ratio is less affected by dust reddening. First, we measure the flux of the local minimum close to 4000 Å and define it as $F\lambda 4000$. Then, we measure the nearest local maximum flux to the blue side of the local minimum and define it as $F\lambda 3950$. We define the flux ratio $FRNi = F\lambda 3950/F\lambda 4000$ and use it as a spectral indicator of Ni abundances. The following equation is used to calculate the local signal-to-noise ratio (SNR) of the spectra:

$$SNR = \frac{F(\lambda)}{STD(\lambda - 30, \lambda + 30)}, \tag{3.5}$$

where λ is the wavelength of interest, $F(\lambda)$ is the flux at the wavelength of interest, and $STD(\lambda - 30, \lambda + 30)$ is the standard deviation of the flux at the pixels between $\lambda - 30$ Å and $\lambda + 30$ Å wavelengths. The SNRs of the local minimum $F\lambda 4000$ and local maximum $F\lambda 3950$ are denoted as SNR_{min} and SNR_{max}, respectively. The uncertainty of $FRNi$ is calculated as

$$\sigma_{FRNi} = \frac{F\lambda 3950}{F\lambda 4000}\sqrt{\frac{1}{SNR_{min}^2} + \frac{1}{SNR_{max}^2}}. \tag{3.6}$$

3.2 Tests of the Neural Networks

3.2.1 Error Estimate

We use two methods to estimate the elemental abundance uncertainty from the NN predictions.

The first method uses the NN-predicted error. In Fig. 3.6, we use the NN to predict the Ni abundance in Zone 3 using the testing dataset and its associated predictive error (σ). When the predictive error is small, we notice that the predicted value is consistent with the true value. With increased predictive error values, the predicted values systematically move away from the truth values. When the predictive errors are larger than 0.4 dex, the predicted values become insensitive to the truth values and take values between −2.0 and −1.5. Therefore, we use $\sigma < 0.4\ dex$ as a criterion to measure the goodness of prediction on the observed spectra.

In the second method, we use the NN-predicted SN ejecta parameters to run a TARDIS model and then compare the model spectrum with the observed spectrum using the MSE value defined in Eq. 3.4. Most of the spectroscopically normal SNe are well fit with MSE ≤ 0.06; on the other hand, fits to peculiar SN 1991bg-like, SN Iax, and some of the SN 1991T or SN 1999aa-like SNe show larger MSEs. This is most likely due to the construction of the model spectral dataset used for training the neural network, which is built based on spectroscopically normal SNe Ia

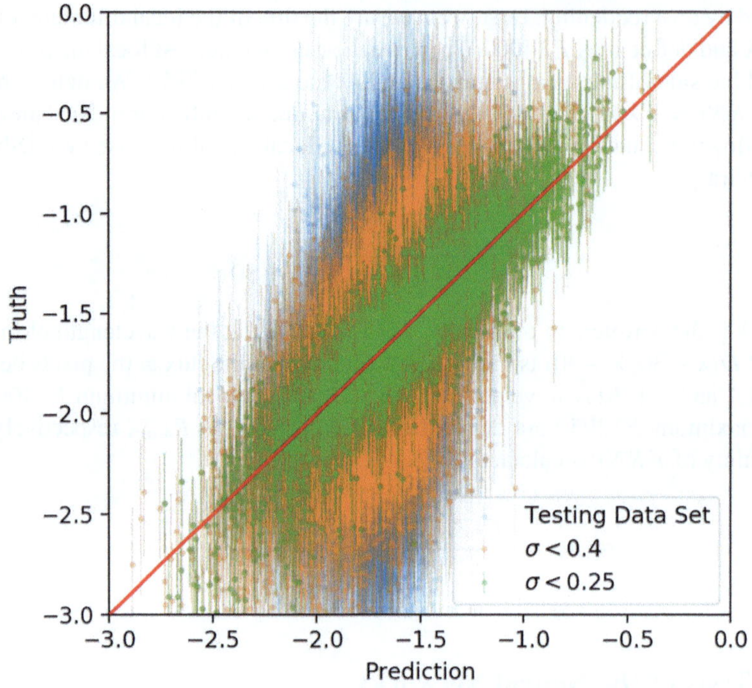

Fig. 3.6 A comparison between the truth values and the predicted values of Ni abundance in Zone 3. The unit of both of the axes is the Ni mass fraction in logarithmic scale ($log_{10}(Ni)$). Data points with large ($\sigma \geq 0.4$), medium ($0.4 > \sigma \geq 0.25$), and small ($\sigma \leq 0.25$) predictive errors are in blue, orange, and green colors, respectively. A reference line (Prediction = Truth) is shown in red. For illustration purposes, we only include 20,000 data points randomly selected from the testing dataset

(see Sect. 3.5 for more discussions). In this chapter, we focus on spectroscopically normal SNe and will focus on models and data with MSE $\leq$ 0.06.

3.2.2 Ni Abundance Sensitivity

In this section, we use the SN ejecta structure predicted from the HST spectra of SN 2011fe at 12.6 days, 19.61 days, 22.89 days, and 28.84 days to simulate four spectral sequences with different Ni abundances in 13,000−17,000 km/s (Zone 3). In these ejecta models, we allow the mass fraction of Ni in Zone 3 to vary from 10^{-4} to 10^{0}, while keeping the relative ratio of other elements in Zone 3 unchanged. Other model parameters, such as the density structures, are kept the same. We employ the NNs to predict the elemental abundances using these simulated spectral sequences. This can test the capability of NN in recovering the assumed model abundances in the context of ejecta models similar to a real SN Ia. Figure 3.7 shows the predicted

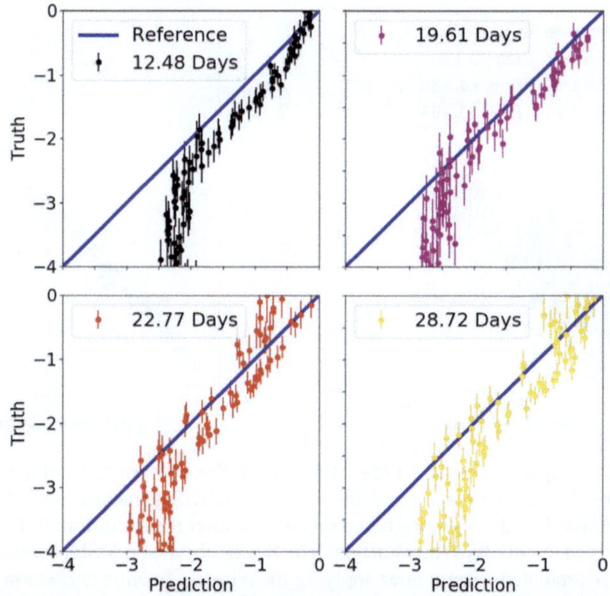

Fig. 3.7 The predicted value compared to the truth value of the neural network with the abundance of ^{56}Ni in Zone 3 as the only varying parameter. The horizontal and vertical axes show the truths and predictions, respectively, of the logarithmic Ni mass fraction $log_{10}(Ni)$ in the velocity range 13,000−17,000 km/s (Zone 3). The spectra used for these predictions are generated from TARDIS based on the ejecta structure derived for SN 2011fe but varying the Ni mass fraction in the velocity range 13,000−17,000 km/s. The blue lines show Prediction = Truth. The times after the explosion of the simulated spectra are labeled in the upper left of each figure

and the true Ni abundance in Zone 3 of the spectral sequences. The typical prediction errors are ∼0.2 dex, and the predicted values are generally consistent with the truth when Ni abundance is larger than $10^{-2.5}$ for the spectra in all four phases. The linear relation is lost when the abundance drops below $\sim 10^{-2.5}$, indicating the sensitivity limit of the models. We surmise the low Ni abundances cannot produce spectral features stronger than the Monte Carlo noise in the simulations and that there may be hidden degeneracies in the model that are still not fully understood.

Moreover, a bias of ∼0.3 dex is observed on the SN 2011fe spectra 12.6 days after the explosion. A possible reason is that electron scattering is still important at Zone 3 at the early epochs, reducing the sensitivity to Ni abundances in the spectral formation. To correct this bias, we use each best-fit simulated spectra as the input to the NNs and its associated best-fit parameters A and B to re-predict the elemental abundances. In the left panel of Fig. 3.8, we compare the re-predicted Ni mass fractions in Zone 3 with those predicted by the observed spectra.

The re-predicted Ni mass fractions are systematically lower than those predicted directly from the observed spectra. The difference is phase-dependent and is shown in three bins of the time after the explosion in Fig. 3.8 (Left). The spectra are split into five groups according to the phase after the explosion to correct this bias. The

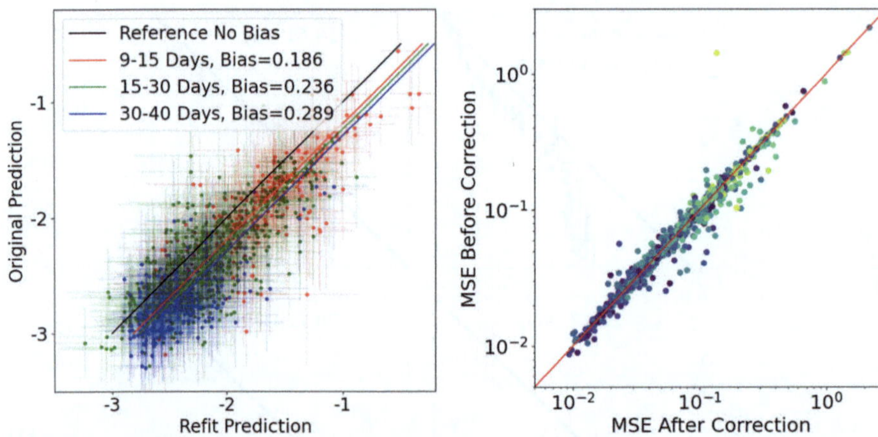

Fig. 3.8 Left: A comparison of the Ni mass fractions in Zone 3 derived from the NN predictions using the observed spectra (Y-axis) and the best-fit TARDIS simulated spectra (X-axis). The spectral data are split into three groups based on the time after the explosion: 9–15 days (red), 15–30 days (green), and 30–40 days (blue). Right: The X-axis shows the MSEs between the observed and the TARDIS simulated spectra after applying the Ni mass fraction correction in Zone 3, and the Y-axis shows the same without this correction. Different colors on the data points indicate the time after the explosion. Purple means early, and yellow means late

time intervals for these five groups are [9,14), [14,19), [19,24), [24,29), [29,34), and [34,40) days. An average bias is calculated for each group. The Ni mass fraction is corrected using the average bias in their corresponding time bin.

We further iterate by calculating simulated spectra using TARDIS with the bias-corrected ejecta structure. The bias-corrected ejecta structure shows insignificant improvements to the MSEs of the fits, as shown in Fig. 3.8 (Right), which compares the MSE differences between the observed and simulated spectra with and without the bias corrections in Zone 3. The absence of a significant difference in the MSE suggests that bias is not the dominant source of our model uncertainties. We emphasize that our spectral library is still far too small to cover the entire parameter space of SNe Ia. The TARDIS code also has its own limitations when applied to model supernova observations. A much larger simulation dataset with more complete coverage of all spectral types of SNe Ia and a fine-tuned NN structure are necessary for future studies. For these reasons, we did not adopt this bias correction in the subsequent studies of this chapter.

3.2.3 Ni Abundances and Data Noise

Most of the observed spectral data have no published errors. This chapter quantifies the SNR using data between 3900 Å and 4100 Å following the method defined in the DER_SNR algorithm [22]. This wavelength range is covered by most of the SNe Ia

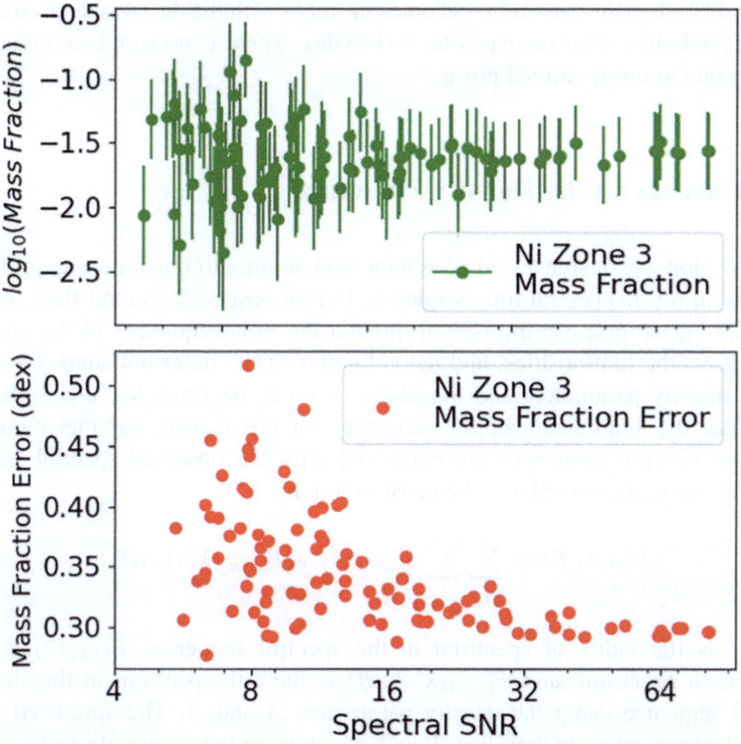

Fig. 3.9 A demonstration of the relation between the spectral SNR and the Ni mass fraction in Zone 3 for the HST spectrum of ASASSN-14lp. The x-axis is the input spectra SNR measured between 3900 Å and 4100 Å, the y-axis in the upper panel is the predicted Ni mass fraction in Zone 3, and the y-axis in the lower panel is the predictive Ni mass fraction error in Zone 3

spectra. The noise is calculated using the following:

$$\sigma = \frac{1.482602}{\sqrt{6}} Med\left(|2F(i) - F(i-2) - F(i+2)|\right), \tag{3.7}$$

where $F(i)$ is the spectral flux at the i-th pixel, and Med calculates the median value over all pixels. The signal is taken as the median flux value.

We use the HST spectrum of ASASSN-14lp at day 14.49, which has an SNR of 81, as an example to discuss the effect of the observational noise on the elemental abundance estimations. The predicted Ni mass fraction at Zone 3 is around $10^{-1.5}$, which is much higher than the $10^{-2.5}$ sensitivity limit according to Sect. 3.2.2. Gaussian noises are applied to the observed spectrum, and the NN is used to estimate the elemental abundances and predictive errors of the Ni element in Zone 3. As shown in Fig. 3.9, for data with SNR $\geq$20, the predicted Ni element mass ratio is stable around $10^{-1.5}$ and the predictive error is around 0.3 dex. This shows an error floor of the abundance estimates and the limitation of the current NNs. The predicted Ni element mass ratio becomes unstable with the decreased spectral SNR, and the predictive error increases to $\sim$0.45 dex. There is no significant bias in the Ni mass due to observational noise.

3.3 Caveats on the Density Structure

In this section, we design a grid-search method, aiming to find a density profile that could fit an SN Ia spectral time sequence. For an observed spectral time sequence of an SN Ia, we first use the NNs to predict the time sequences of the elemental abundances, the luminosities, and the velocities of the inner boundary for a given grid of density parameters and the phase of each spectrum. We use TARDIS to recalculate the simulated spectral sequences for this density parameter grid. The simulated spectral sequences are compared with the observed spectral sequence using the squared error (SE) as the merit function:

$$SE(A, B) = \sum_i \sum_\lambda (F_{i,obs}(\lambda) - F_{i,sim}(\lambda, A, B))^2, \tag{3.8}$$

where i is the index of spectrum in the spectral sequence, $F_{i,obs}(\lambda)$ is the i-th observed spectrum, and $F_{i,sim}(\lambda, A, B)$ is the i-th spectrum in the simulated spectral sequence using the density parameters A and B. The simulated spectra are masked according to their wavelength coverage of the observed spectra with the missing wavelength regions set to have zero weights and normalized by dividing the spectra by the average pixel value. The observed spectra are resampled with the wavelength grid of the simulated spectra and are normalized by their average pixel values.

The density profile equation 3.1 has only two free parameters A and B to be determined. However, finding the best A and B parameters that minimize Eq. 3.8 in a typical 19×21 grid space still involves a large number of spectral model computations. We have applied the following method for further improvement. The method has five iterations, and each iteration calculates the spectral sequence in a different sub-grid. In the first iteration, the sub-grid is set to be $A = [0.5, 1.1, 1.7]$, $B = [0.5, 1.1, 1.7]$. In the second iteration, the sub-grid is specified as $A = [a - 0.2, a, a + 0.2]$, $B = [b - 0.2, b, b + 0.2]$, where a and b are the parameters set with the smallest SE from the first iteration. In the third iteration, the sub-grid is replaced by $A = [a - 0.2, a, a + 0.2]$, $B = [b - 0.2, b, b + 0.2]$ with a and b being the values that give the smallest SE from the second iteration. In the fourth iteration, the sub-grid is refined to $A = [a-0.1, a, a+0.1]$, $B = [b-0.1, b, b+0.1]$, where a and b are the parameters set with the smallest SE from the third iteration. In the fifth iteration, the sub-grid is refined to $A = [a-0.1, a, a+0.1]$, $B = [b-0.1, b, b+0.1]$, where a and b are the parameters set with the smallest SE from the third iteration. Finally, the parameter set with the smallest SE among all the sub-grids will be adopted to be the density parameter of the supernova.

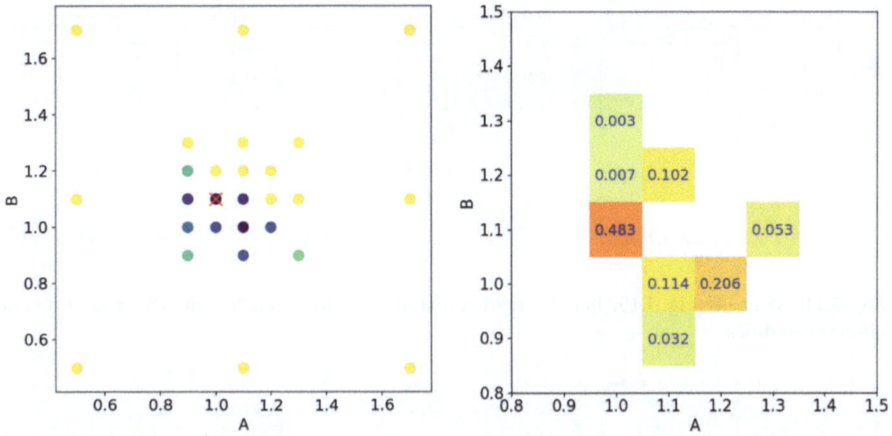

Fig. 3.10 Left panel: The grid-search result of SN 2011fe. X-axis is the density parameter A, and Y-axis is the density parameter B. Yellow dots indicate large SE values, and purple dots indicate small SE values. The red cross marks the grid point with the smallest SE values. Right panel: The histogram of the density parameters for SN 2011fe. Area with high probability is marked in orange color, and area with low probability is marked in green color. The proportion of each grid to the whole distribution is marked in blue text

For a well-observed SN with $\sim$30 spectra (e.g., SN 2005cf), the total computation time could still exceed 1000 CPU hours using this grid-search method. Therefore, we limit the number of observed spectra used in the grid search to about 10 based on the quality of the spectral data. For the well-observed SNe, we only select the spectra from ground-based telescopes with apertures larger than 3 meters in diameter and the Hubble Space Telescope (HST). The spectra are visually inspected to eliminate the data with obvious abnormal flux calibrations.

We use 11 spectra of SN 2011fe between 9 days and 40 days after the explosion to explore the best-fit density parameters, the observation time, and the instruments are shown in Fig. 3.4. The left panel of Fig. 3.10 shows the SE values (Eq. 3.8) of several selected grid points in the density parameter space. The number of TARDIS models to be calculated is $(9+8+0+8+5) \times N_{obs}$, where 9, 8, 0, 8, and 5 are the number of grid points in the five iterations described above, and N_{obs} is the number of observed spectra. The grid points that are calculated in the previous iterations will not be calculated again; therefore, the 2nd, 4th, and 5th iterations have numbers of grid points smaller than 9. Because the best point after the second iteration is $A = 1.1$, $B = 1.1$, the sub-grid in the third iteration overlaps with the searched grid points, and no calculation is executed in the third iteration. Finally, the best-fit density parameters A and B are found to be 1.0 and 1.1 for SN 2011fe.

The simulated spectral time sequence using this fixed density structure is shown as the gray lines in Fig. 3.4.

In order to test the stability of this density estimate method, we use SN 2011fe as example and apply bootstrap resampling method. In each of the iteration, we randomly delete half of the observed spectra and then use the remaining spectra to

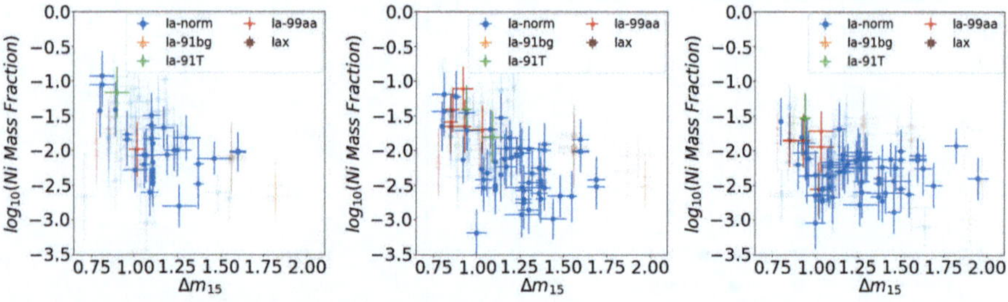

Fig. 3.11 Same as Fig. 3.15, but the density parameters are fixed for an SN Ia at different observation times

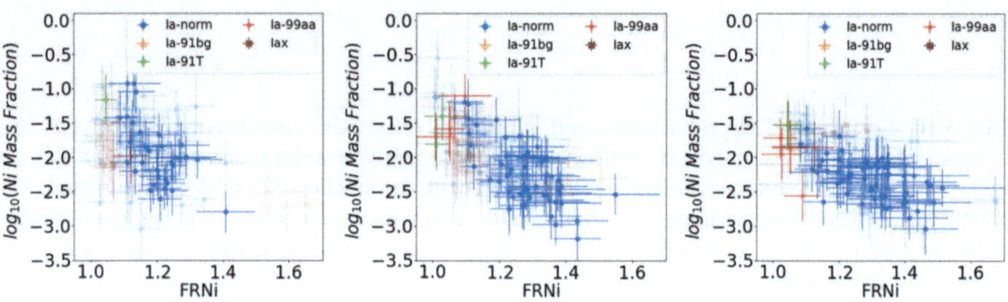

Fig. 3.12 Same as Fig. 3.16, but the density parameters are fixed for an SN Ia at different observation times

estimate the density parameters. We run 1000 iterations and then draw a histogram of the estimated density parameter distributions, as shown in Fig. 3.10. The result on SN 2011fe shows that about 45% of the estimated density parameters are the same as the estimated density parameters using all of the observed spectra. Moreover, the density distribution in the SN 2011fe result shows roughly two clusters.

In Figs. 3.11 and 3.12, we compare the estimated Ni element mass ratio in Zone 3 with the light curve parameter Δm_{15} and the spectral feature $FRNi$ value again. The density parameters are fixed for an SN Ia at different observation times to obtain the minimal SE value between the simulated spectral time sequence and the observed spectral time sequence. We notice the correlations between the Ni mass ratio and Δm_{15}, between the Ni mass ratio and $FRNi$, are retained with the change of density parameters. Moreover, we observe many spectral fitting results provide an MSE larger than 0.06 criterion, resulting in more transparent data points in the figures.

To conclude, using a two-parameter density model in Eq. 3.1 could be an oversimplification to the actual SN Ia density structures, especially for the SNe Ia with multi-epoch observations such as SN 2011fe. A density profile with more parameters or a three-dimensional density structure could be used in the future SNe Ia structure modeling researches, in order to provide a more self-consistent ejecta structure.

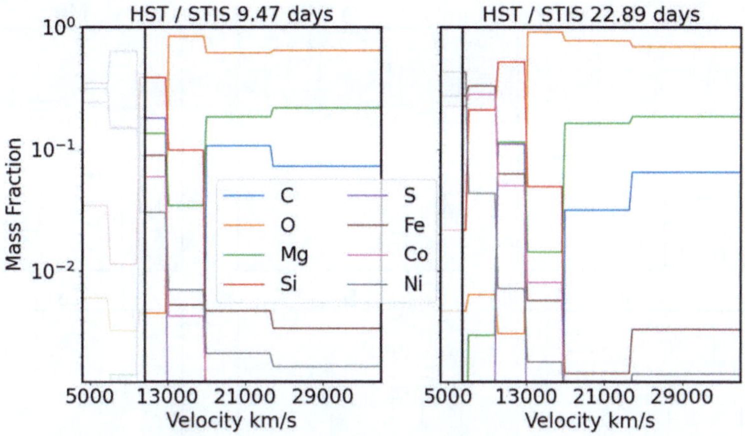

Fig. 3.13 The elemental abundances predicted from the SN 2011fe spectra on days 9.47 (left) and 22.89 (right). The thick vertical black lines mark the velocities of the TARDIS inner boundary, which are also one of the parameters predicted by the NNs

3.4 SN 2011fe as an Example

This section uses SN 2011fe to show the deep-learning-based pipeline's spectral fitting performance and elemental abundance accuracy, using the well-observed SN 2011fe as an example.

Figure 3.13 shows the elemental abundances derived from the observed spectra of SN 2011fe on days 9.47 and 22.89. Only the chemical structure above the photosphere can be characterized with the spectra at the photospheric phase. The elemental abundances derived from these two spectra are highly consistent at Zones above 10,000 km/sec despite of the rapid evolution of the spectral features during the two epochs (Fig. 3.4).

The abundance structure of SN 2011fe we derived can be compared with the $\rho-11fe$ model [23, 24] emanated from spectral modeling and the DDT_2013_N100 model [6] calculated from hydrodynamic and nucleosynthesis simulations, as shown in Fig. 3.14. The elemental abundances derived using data 22.89 days after the explosion are used for these comparisons because the observed spectrum covers a broad wavelength range of 2000 to 10,000 Å, and the observed spectrum is well fitted by the TARDIS simulated spectrum (see Fig. 3.4). For the DDT_2013_N100 model (downloadable at HESMA[1]), specific isotopes are available, while for the $\rho-11fe$ model, we assume all the Ni to be ^{56}Ni initially and calculate the elemental abundances at day 22.89 by propagating the nuclear decay processes of radioactive elements.

[1] https://hesma.h-its.org/

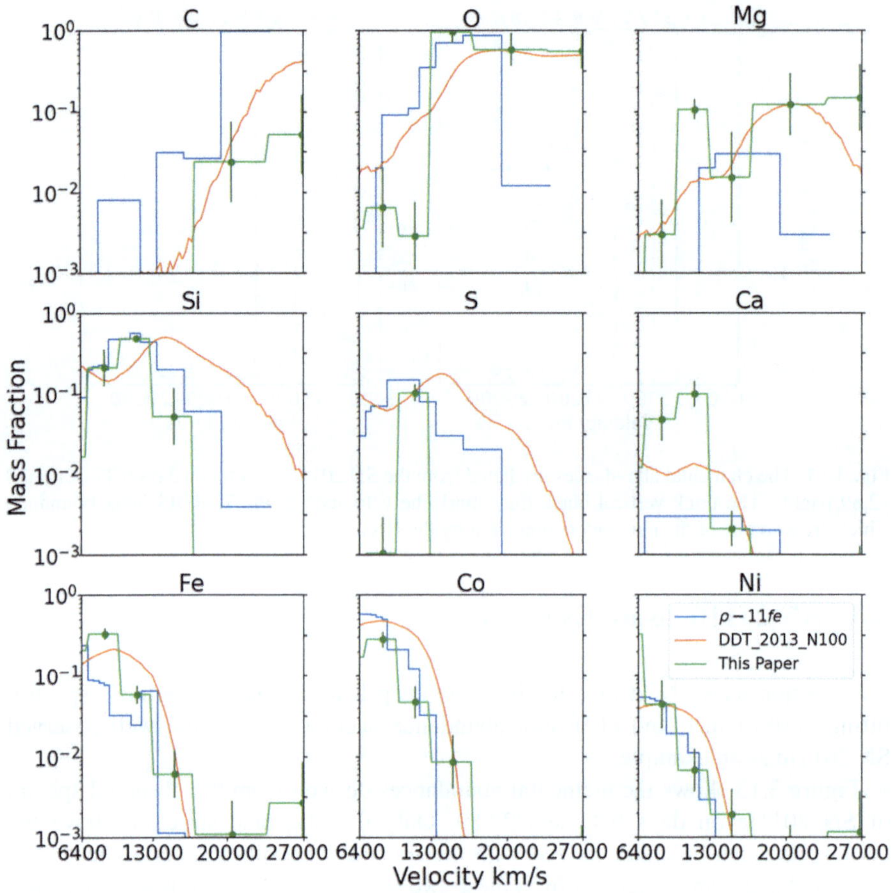

Fig. 3.14 Comparisons of the ejecta structure models of SN 2011fe. The figure shows the $\rho - 11fe$ model (blue lines), the DDT_2013_N100 model (orange lines), and the chemical structure model of this study using spectral data at day 22.89 (green lines). The chemical elements are identified in the title of each panel. The NN-predicted errors are shown as green error bars

The three ejecta structures are remarkably consistent in the Fe, Co, and Ni distribution (Fig. 3.14, bottom panels). A sharp drop-off is seen at velocities above 20,000 km/sec. Our model is also broadly consistent with DDT_2013_N100 in other elements such as C, O, Mg, and Ca, but significant discrepancies are found for Si and S where our NN predicts much less Si and S at velocities above 15,000 km/sec than DDT_2013_N100. The Ca abundance from this study is much higher than the other two models at velocities below 13,000 km/sec, but note that the Ca features are strongly affected by NLTE effect. The C and O abundance is consistent above Zone 3 among the three models, confirming the presence of unburned materials at the outermost layers of the ejecta. Overall, the agreement between the NN-derived chemical structure and DDT_2013_N100 is remarkable, supporting the notion that SN 2011fe is the result of the explosion of a Chandrasekhar mass white dwarf.

3.5 Applications of the Method and the Results

We searched all the available SN Ia spectra from WISeREP [25] and kaepora [18] and found 1087 spectra from 124 SNe Ia, which were observed between −10 days and 20 days relative to the B-band maximum time with the rest frame wavelength covering 3800−7000 Å and the spectral SNR (as defined in Sect. 3.2.3) larger than 20. This forms the dataset for our analysis to apply the NNs to a sample of well-observed SNe. We also collected the Δm_{15} parameters of these SNe from kaepora [18] or calculated the Δm_{15} parameters using the light curves from kaepora [18] or the Open Supernova Catalog.[2] All the spectra are converted into rest frame and corrected for dust extinction effects using the data from kaepora [18]. The times after the explosion of these SNe are calculated assuming the B-band maximum time is 19 days after the explosion for simplicity. Moreover, because the inner boundary velocity of TARDIS falls in Zone 2 for most of the SN models at the epochs between − 10 and 0 days from the B-band maximum, we only use the Ni abundance in Zone 3, which is less affected by the location of the TARDIS inner boundary for the early-phase spectral models, as measures of Ni abundances when comparing the ejecta models to the observations.

Figure 3.15 shows the relation between Δm_{15} and the Ni abundance in Zone 3 of the SN sample. The dataset contains 416 spectra from 124 SNe observed between 9 and 18 days after the explosion. The figure shows the correlations between Δm_{15} and the Ni mass fraction in Zone 3 for the spectra between [9, 12), [12,15), and [15, 18) days after the explosion. The correlation is much weaker during 15 and 18 days after the explosion. The Ni abundances are significantly lower than in earlier epochs, and for most of the SNe, the values are close to the detection limit. The Ia-99aa and Ia-91T sub-types are found in the regions with more significant Ni mass fraction and smaller Δm_{15}. The Ia-91bg subtype shows large Δm_{15} and low Ni mass fraction, consistent with being underluminous due to the deficiency of ^{56}Ni in the ejecta as the energy source. The Iax sub-type SNe is comparatively isolated from the rest of the SNe Ia, showing large Δm_{15} and large Ni mass fraction. Moreover, the Ni abundance in Zone 3 does not show a significant decrease for type Iax SNe. If confirmed by observations and models, this may indicate that the Ni in Type Iax consists primarily of stable isotopes of Ni. SN 2007hj and SN 2005dv are classified as normal SNe Ia [20] but with large Δm_{15} values in the right panel of Fig. 3.15.

The relations between the Ni abundances in Zone 3 and the spectral line ratio *FRNi* are shown in Fig. 3.16 for the three epoch bins as in Fig. 3.15. *FRNi* shows a linear correlation to the logarithmic Ni abundance in Zone 3 $log_{10}(Ni)$, especially

[2] https://sne.space/

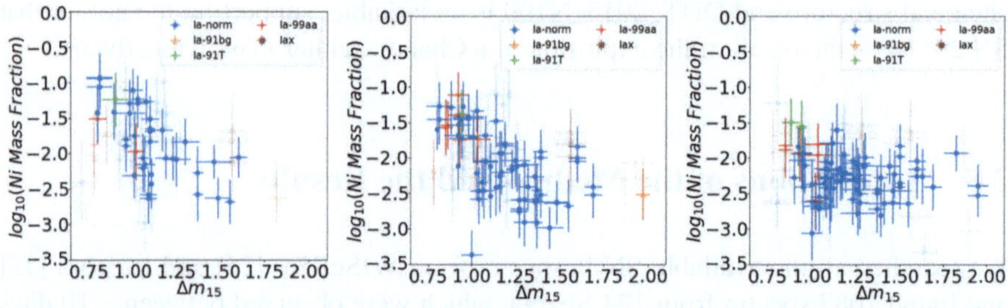

Fig. 3.15 The relation between Δm_{15} and the Ni abundance in Zone 3 at different time bins. Left panel: between 9 and 12 days after the explosion. Middle panel: between 12 and 15 days after the explosion. Right panel: between 15 and 18 days after the explosion. y-axis is the Ni mass fraction in Zone 3 in the logarithmic scale. Spectroscopically normal SNe Ia are marked with blue solid circles, Ia-91bg-like SNe are marked with orange hollow triangles, Ia-91T-like SNe are marked with green diamonds, Ia-99aa-like SNe are marked with red solid stars, and Iax SNe are marked with brown squares. Data points with large errors on the y-axis ($\sigma > 0.4$) or large spectral fitting error (MSE > 0.06) are transparent

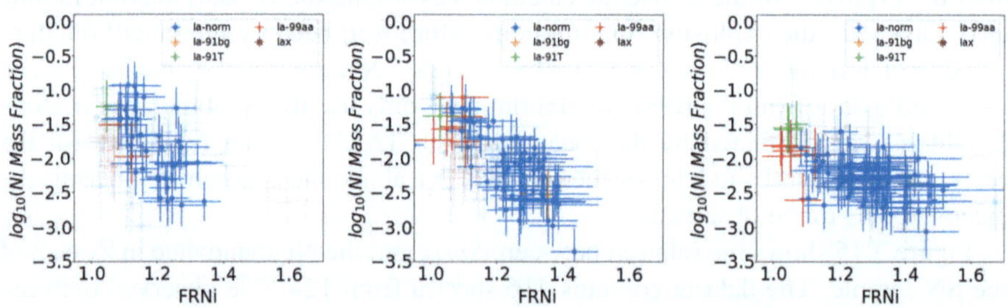

Fig. 3.16 Same as Fig. 3.15 but for the relation between the spectral line ratio ($FRNi$) measured from the observed spectra of SNe and the Ni abundance in Zone 3 determined by applying the NN models to the observed data

for the spectra between 12 and 15 days after the explosion. Such a linear correlation is weaker for the spectra between 15 and 18 days after the explosion, probably because the Ni abundances have decreased below the detection limit for most of the SNe. The Ia-99aa and Ia-91T sub-type SNe show higher Ni abundances and lower *FRNi* parameters than spectroscopically normal SNe Ia. The Ia-91bg sub-type SNe shows lower Ni abundances and larger *FRNi*.

In Fig. 3.17, we show the relation between the spectral line ratio (*FRNi*) measured from the observed spectra and the Δm_{15} values. A linear relation exists between the *FRNi* values and the Δm_{15} values for most of the subtypes of SNe Ia at different phases between 9 and 18 days after the explosion. The Ia-91bg sub-type shows higher *FRNi* and Δm_{15} values than normal SNe Ia. The Ia-91T and Ia-99aa sub-types show the lowest Δm_{15} and *FRNi* compared to other sub-types of SNe Ia. The Iax sub-type shows low *FRNi* and high Δm_{15} value, which is slightly off from the correlation defined by the other SN Ia subtypes.

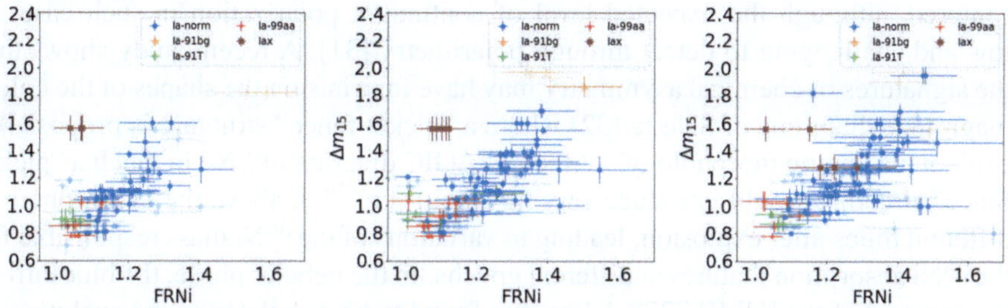

Fig. 3.17 Same as Fig. 3.15 but for the relation between the spectral line ratio ($FRNi$) measured from the observed spectra of SNe and the Δm_{15} values

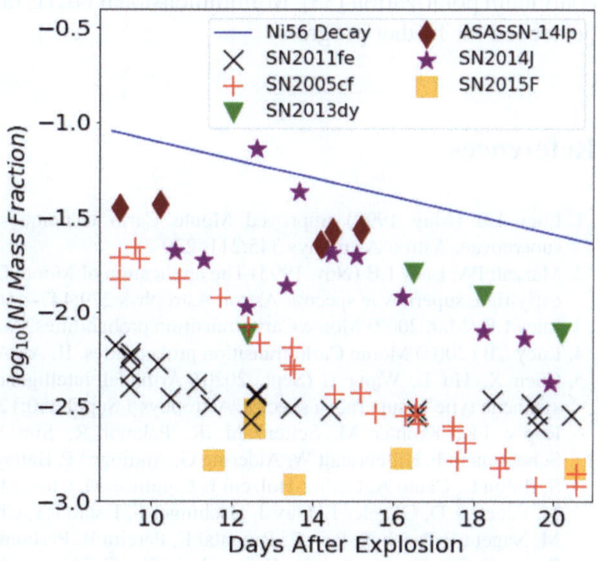

Fig. 3.18 The Ni abundance evolution with time after SN explosion for different SNe Ia. The y-axis is the Ni mass fraction in Zone 3 on a logarithmic scale. The ultramarine line is the theoretical ^{56}Ni abundance changing with time. Six well-observed type Ia-normal SNe: SN2011fe (black cross), SN2005cf (red plus), SN2013dy (green triangle), ASASSN-14lp (brown diamond), SN2014J (purple star), and SN2015F (yellow square) are shown with different symbols

Figure 3.18 shows the time evolution of the Ni abundance in Zone 3 for a few best observed SNe: SN 2011fe, SN 2005cf, SN 2013dy, ASASSN-14lp, SN 2014J, and SN 2015F. The decay rates, as shown in Fig. 3.18, deserve further remarks. While the values of Ni mass fraction of SN 2013dy, ASASSN-14lp, and SN 2014J appear to be consistent with the half-life time of ^{56}Ni, that of SN 2011fe is considerably flatter, and that of SN 2005cf is much steeper than given by the radioactive decay rate. One possibility is that this is intrinsic to the approximations adopted in the code TARDIS, which only utilizes approximated NLTE processes. On the other hand, taking the numbers literally may indicate the inadequacy of the spherical geometry we have adopted. During these observations, even for the fixed velocity zones, the lines are formed in different regions as the photosphere recedes considerably to deeper layers at later times. Chemical inhomogeneities are found to be common in literally all SNe Ia ejecta from polarimetry observations [26–30]. Theoretical models suggest that the chemical structure of SNe Ia may be considerably off-centered, although the expected level of continuum polarization in such cases is low and challenging to detect through polarimetry [31]. A recent study shows that the signatures of chemical asymmetry may have imprints on the shapes of the color-magnitude diagrams of SNe Ia [32] where a "picket fence" structure is proposed to explain the "bump morphology" of the CMAGIC diagram of SNe Ia. Such a "picket fence" structure would introduce varying amounts of ^{56}Ni above the photosphere at different times after explosion, leading to variations of the ^{56}Ni mass responsible for the ^{56}Ni absorption features at different epochs. In the nebular phase, the blueshift of the centroid of the [Ni II] 7378 Å line was found to be correlated to the evolution of the velocity of the P-Cygni absorption of the Si II 6355 Å line, and it was proposed that global asymmetry may be expected in all SNe Ia despite the lack of significant continuum polarization [33]. Multidimensional NLTE radiative transfer models may be needed for further progress.

References

1. Lucy LB (May 1999) Improved Monte Carlo techniques for the spectral synthesis of supernovae. Astron Astrophys 345:211–220
2. Mazzali PA, Lucy LB (Nov. 1993) The application of Monte Carlo methods to the synthesis of early-time supernovae spectra. Astron Astrophys 279:447–456
3. Lucy LB (Mar. 2002) Monte Carlo transition probabilities. Astron Astrophys 384:725–735
4. Lucy LB (2003) Monte Carlo transition probabilities. II. A&A 403(1):261–275
5. Chen X, Hu L, Wang L (Sept. 2020) Artificial intelligence-assisted inversion (AIAI) of synthetic type Ia supernova spectra. Astrophys J Suppl 250:12
6. Röpke FK, Kromer M, Seitenzahl IR, Pakmor R, Sim SA, Taubenberger S, Ciaraldi-Schoolmann F, Hillebrandt W, Aldering G, Antilogus P, Baltay C, Benitez-Herrera S, Bongard S, Buton C, Canto A, Cellier-Holzem F, Childress M, Chotard N, Copin Y, Fakhouri HK, Fink M, Fouchez D, Gangler E, Guy J, Hachinger S, Hsiao EY, Chen J, Kerschhaggl M, Kowalski M, Nugent P, Paech K, Pain R, Pecontal E, Pereira R, Perlmutter S, Rabinowitz D, Rigault M, Runge K, Saunders C, Smadja G, Suzuki N, Tao C, Thomas RC, Tilquin A, Wu C May (2012) Constraining type Ia supernova models: SN 2011fe as a test case. Astrophys J Lett 750:L19
7. Kromer M, Fink M, Stanishev V, Taubenberger S, Ciaraldi-Schoolman F, Pakmor R, Röpke FK, Ruiter AJ, Seitenzahl IR, Sim SA, Blanc G, Elias-Rosa N, Hillebrandt W (Mar. 2013) 3D deflagration simulations leaving bound remnants: a model for 2002cx-like type Ia supernovae. Mon Not R Astron Soc 429:2287–2297
8. Pakmor R, Kromer M, Taubenberger S, Sim SA, Röpke FK, Hillebrandt W (Mar. 2012) Normal type Ia supernovae from violent mergers of white dwarf binaries. Astrophys J Lett 747:L10
9. Sim SA, Fink M, Kromer M, Röpke FK, Ruiter AJ, Hillebrandt W (Mar. 2012) 2D simulations of the double-detonation model for thermonuclear transients from low-mass carbon-oxygen white dwarfs. Mon Not R Astron Soc 420:3003–3016
10. Hsiao EY, Conley A, Howell DA, Sullivan M, Pritchet CJ, Carlberg RG, Nugent PE, Phillips MM (2007) K-corrections and spectral templates of type Ia supernovae. Astrophys J 663:1187–1200
11. Hu L, Chen X, Wang L (May 2022) Spectroscopic studies of type Ia supernovae using LSTM neural networks. Astrophys J 930:70
12. LeCun Y, Bengio Y, Hinton G (2015) Deep learning. Nature 521:436–444
13. Lakshminarayanan B, Pritzel A, Blundell C (Dec. 2016) Simple and scalable predictive uncertainty estimation using deep ensembles. arXiv e-prints, arXiv:1612.01474

14. Krizhevsky A, Sutskever I, Hinton GE (2012) ImageNet classification with deep convolutional neural networks. In: Pereira F, Burges CJC, Bottou L, Weinberger KQ, (eds) Advances in neural information processing systems 25. Curran Associates, Inc., Red Hook, pp 1097–1105
15. Simonyan K, Zisserman A (Sep 2014) Very deep convolutional networks for large-scale image recognition. arXiv e-prints, arXiv:1409.1556
16. Abdi M, Nahavandi S (2016) Multi-residual networks. CoRR, abs/1609.05672
17. Kingma DP, Ba J (2014) Adam: a method for stochastic optimization. CoRR, abs/1412.6980
18. Siebert MR, Foley RJ, Jones DO, Angulo R, Davis K, Duarte A, Strasburger E, Conlon M, Kazmi N, Nishimoto R, Schubert M, Sun L, Tippens R (2019) Investigating the diversity of Type Ia supernova spectra with the open-source relational data base kaepora. Mon Not R Astron Soc 486:5785–5808
19. Blondin S, Blinnikov S, Callan FP, Collins CE, Dessart L, Even W, Flörs A, Fullard AG, Hillier DJ, Jerkstrand A, Kasen D, Katz B, Kerzendorf W, Kozyreva A, O'Brien J, Pássaro EA, Roth N, Shen KJ, Shingles L, Sim SA, Singhal J, Smith IG, Sorokina E, Utrobin VP, Vogl C, Williamson M, Wollaeger R, Woosley SE, Wygoda N (Dec. 2022) StaNdaRT: a repository of standardised test models and outputs for supernova radiative transfer. Astron Astrophys 668:A163
20. Silverman JM, Ganeshalingam M, Li W, Filippenko AV (2012) Berkeley supernova Ia program—III. Spectra near maximum brightness improve the accuracy of derived distances to type Ia supernovae. Mon Not R Astron Soc 425:1889–1916
21. Walker ES, Hook IM, Sullivan M, Howell DA, Astier P, Balland C, Basa S, Bronder TJ, Carlberg R, Conley A, Fouchez D, Guy J, Hardin D, Pain R, Perrett K, Pritchet C, Regnault N, Rich J, Aldering G, Fakhouri HK, Kronborg T, Palanque-Delabrouille N, Perlmutter S, Ruhlmann-Kleider V, Zhang T (2010) Supernova Legacy Survey: using spectral signatures to improve type Ia supernovae as distance indicators. Mon Not R Astron Soc 410:1262–1282
22. Stoehr F, White R, Smith M, Kamp I, Thompson R, Durand D, Freudling W, Fraquelli D, Haase J, Hook R, Kimball T, Kümmel M, Levay K, Lombardi M, Micol A, Rogers T (Aug. 2008) DER_SNR: a simple & general spectroscopic signal-to-noise measurement algorithm. In: Argyle RW, Bunclark PS, Lewis JR (eds) Astronomical data analysis software and systems XVII vol 394. Astronomical Society of the Pacific Conference Series, p 505
23. Mazzali PA, Sullivan M, Hachinger S, Ellis RS, Nugent PE, Howell DA, Gal-Yam A, Maguire K, Cooke J, Thomas R, Nomoto K, Walker ES (Apr. 2014) Hubble space telescope spectra of the type Ia supernova SN 2011fe: a tail of low-density, high-velocity material with Z $Z_{\odot}$. Mon Not R Astron Soc 439:1959–1979
24. Heringer E, van Kerkwijk MH, Sim SA, Kerzendorf WE (Sept. 2017) Spectral sequences of type Ia supernovae. I. Connecting normal and subluminous SNe Ia and the presence of unburned carbon. Astrophys J 846:15
25. Yaron O, Gal-Yam A (July 2012) WISeREP - an interactive supernova data repository. Publ Astron Soc Pacif 124:668
26. Wang L, Baade D, Höflich P, Khokhlov A, Wheeler JC, Kasen D, Nugent PE, Perlmutter S, Fransson C, Lundqvist P (July 2003) Spectropolarimetry of SN 2001el in NGC 1448: asphericity of a normal type Ia supernova. Astrophys J 591:1110–1128
27. Wang L, Wheeler JC (Sept. 2008) Spectropolarimetry of supernovae. Ann Rev Astron Astrophys 46:433–474
28. Cikota A, Patat F, Wang L, Wheeler JC, Bulla M, Baade D, Höflich P, Cikota S, Clocchiatti A, Maund JR, Stevance HF, Yang Y (Nov. 2019) Linear spectropolarimetry of 35 Type Ia supernovae with VLT/FORS: an analysis of the Si II line polarization. Mon Not R Astron Soc 490:578–599
29. Yang Y, Hoeflich P, Baade D, Maund JR, Wang L, Brown PJ, Stevance HF, Arcavi I, Burke J, Cikota A, Clocchiatti A, Gal-Yam A, Graham ML, Hiramatsu D, Hosseinzadeh G, Howell DA, Jha SW, McCully C, Patat F, Sand DJ, Schulze S, Spyromilio J, Valenti S, Vinkó J, Wang X, Wheeler JC, Yaron O, Zhang J (Oct. 2020) The young and nearby normal type Ia supernova 2018gv: UV-optical observations and the earliest spectropolarimetry. Astrophys J 902:46

30. Yang Y, Yan H, Wang L, Wheeler JC, Baade D, Isaacson H, Cikota A, Maund JR, Hoeflich P, Patat F, Giacalone S, Rice M, Tyler DB, Mishra D, Ashall C, Brink TG, Filippenko AV, Galbany L, Patra KC, Shahbandeh M, Vasylyev SS, Vinkó J (Oct 2022) Spectropolarimetry of the thermonuclear supernova SN 2021rhu: high calcium polarization 79 days after peak luminosity. Astrophys J 939:18
31. Hoeflich P, Yang Y, Baade D, Cikota A, Maund JR, Mishra D, Patat F, Patra KC, Wang L, Wheeler JC, Filippenko AV, Gal-Yam A, Schulze S (Mar. 2023) The core normal Type Ia supernova 2019np - an overall spherical explosion with an aspherical surface layer and an aspherical ^{56}Ni core. Mon Not R Astron Soc 520:560–582
32. Aldoroty L, Wang L, Hoeflich P, Yang J, Suntzeff N, Aldering G, Antilogus P, Aragon C, Bailey S, Baltay C, Bongard S, Boone K, Buton C, Copin Y, Dixon S, Fouchez D, Gangler E, Gupta R, Hayden B, Karmen M, Kim AG, Kowalski M, Küsters D, Léget PF, Mondon F, Nordin J, Pain R, Pecontal E, Pereira R, Perlmutter S, Ponder KA, Rabinowitz D, Rigault M, Rubin D, Runge K, Saunders C, Smadja G, Suzuki N, Tao C, Thomas RC, Vincenzi M (Oct. 2022) Bump morphology of the CMAGIC diagram. arXiv e-prints, arXiv:2210.06708
33. Maeda K, Benetti S, Stritzinger M, Röpke FK, Folatelli G, Sollerman J, Taubenberger S, Nomoto K, Leloudas G, Hamuy M, Tanaka M, Mazzali PA, Elias-Rosa N (July 2010) An asymmetric explosion as the origin of spectral evolution diversity in type Ia supernovae. Nature 466:82–85

Chapter 4
An Integral-Based Technique (IBT) to Accelerate the Monte Carlo Radiative Transfer Computation for Supernovae

Abstract We present an integral-based technique (IBT) algorithm to accelerate supernova (SN) radiative transfer calculations. The algorithm utilizes "integral packets," which are calculated by the path integral of the Monte Carlo energy packets, to synthesize the observed spectropolarimetric signal at a given viewing direction in a 3-D time-dependent radiative transfer program. Compared to the event-based technique (EBT) proposed by Bulla et al. (Mon Not R Astron Soc 450:967–981;2015) our algorithm significantly reduces the computation time and increases the Monte Carlo signal-to-noise ratio. Using a 1-D spherical symmetric type Ia supernova (SN Ia) ejecta model DDC10 and its derived 3-D model, the IBT algorithm has successfully passed the verification of spherical symmetry and cross comparison on a 3-D SN model with direct counting technique (DCT) and EBT. Notably, with our algorithm implemented in the 3-D Monte Carlo radiative transfer code SEDONA, the computation time is faster than EBT by a factor of 10−30, and the signal-to-noise (S/N) ratio is better by a factor of 1.5−3, with the same number of Monte Carlo quanta.

4.1 Spectrum Extraction

In the SEDONA program, the optical energy packets (OP) are created to represent a collection of photons with the same frequency, coordinate, and propagating direction. The OPs undergo physical events including Thomson scattering, bound-bound transition, bound-free transition, and free-free transition in the SN plasma to change the frequency and direction, finally escaping the SN.

X. Chen, *Deep Learning on Type Ia Supernovae*, Springer Theses,
https://doi.org/10.1007/978-3-032-13032-7_4

The polarization information of the OP is stored as a dimensionless Stokes vector:

$$\boldsymbol{s}_{OP} = \begin{pmatrix} 1 \\ q \\ u \\ v \end{pmatrix}, \tag{4.1}$$

and the Stokes vector can be expressed as

$$\boldsymbol{S}_{OP} = \begin{pmatrix} I \\ Q \\ U \\ V \end{pmatrix} = E_{OP}\boldsymbol{s}_{OP}, \tag{4.2}$$

where E_{OP} is the energy of the OP.

The OPs are created from the gamma-ray scattering and energy deposition and thus have no initial polarization. During the propagation of an OP, the linear polarization state changes with the Thomson scattering process following the Rayleigh scattering rules (Section 1.17, equation 217 of [1]). In bound-bound transition, bound-free transition, and free-free transition, the OPs are reassigned an unpolarized Stokes vector. The current version of SEDONA did not include magnetic field calculation, and thus the V parameter in the Stokes vector, which represents the circular polarization, is set to zero.

4.1.1 Direct Counting Technique

A straightforward method of spectral synthesis in MCRT is the direct counting technique (DCT), which counts the energy of the OPs to sum up the spectral flux:

$$\begin{pmatrix} \mathcal{I} \\ \mathcal{Q} \\ \mathcal{U} \\ \mathcal{V} \end{pmatrix} = \frac{1}{4\pi r^2 \Delta t \Delta \nu} \sum_i E_{OP,i}\boldsymbol{s}_{OP,i}, \tag{4.3}$$

where $\boldsymbol{s}_{OP,i}$ is dimensionless Stokes vector of the i-th OP, $E_{OP,i}$ is the rest frame energy of the i-th OP, Δt is the size of the arrival time bin, $\Delta \nu$ is the size of the frequency bin, and r is the distance to the SN center. The summation is over all the OPs in the arrival time interval $[t - \Delta t/2, t + \Delta t/2)$, the frequency interval $[\nu - \Delta \nu/2, \nu + \Delta \nu/2)$, and the viewing direction interval $[\varphi - \Delta\varphi/2, \varphi + \Delta\varphi/2)$; $[\theta - \Delta\theta/2, \theta + \Delta\theta/2)$.

The S/N of the simulation spectra depends on the number of OPs and the choice of the bins of arrival time, frequency, and viewing direction. Therefore, a refined viewing direction bin aiming to a high angular resolution could result in the lack of OPs per bin and a low S/N.

DCT is implemented in the SEDONA code by default.

4.1.2 Virtual Packets

Lucy [2] discussed the usage of formal integral in calculating a high S/N spectra based on a MCRT simulation. Further implementations are introduced in the multidimensional simulations of active galactic nuclei (e.g., [3]) and stellar accretion disk winds (e.g., [4]). The concept of "virtual packet" (VP) in the SN simulation context is introduced by Kerzendorf and Sim [5] and is further implemented in the 3-D simulation by Bulla et al. [6], for the purpose of increasing the signal-to-noise ratio (S/N) in the MCRT simulation. Generally, the VPs follow the rules:

- A viewing direction is specified before the MCRT simulation.
- The VPs are created in the SN plasma to represent the emitted and scattered photons at the corresponding volume-time-frequency bin pointing to the viewing direction.
- The VPs propagate with the speed of light toward the viewing direction, and the energy of each VP is reduced by the opacity over the trajectory.
- Finally, the escaped VPs are used for spectral synthesis.

Among the two VP methods proposed in [6], the event-based technique (EBT) provides a good balance between the computation time and the result S/N.

In the EBT calculation, when an OP has undergone a physical event, a VP is created at the same coordinate and time with the same co-moving frame frequency. If the physical event is Thomson scattering, then the co-moving frame Stokes vector of the VP follows the Rayleigh scattering rule[1] [1]:

$$\bar{\boldsymbol{S}}_{VP} = \bar{\boldsymbol{S}}_{OP}\boldsymbol{P}(\bar{\theta}_{OP}, \bar{\varphi}_{OP}, \bar{\theta}, \bar{\varphi}) \, , \tag{4.4}$$

where $\bar{\boldsymbol{S}}_{OP}$ is the Stokes vector of the OP in the co-moving frame, $(\bar{\theta}_{OP}, \bar{\varphi}_{OP})$ is the OP propagating direction, $(\bar{\theta}, \bar{\varphi})$, is the viewing direction in the co-moving frame, and $\boldsymbol{P}$ is the combination of the rotation matrix and the scattering matrix (See Equation 10 in [6]). A detailed expression of the $\boldsymbol{P}$ matrix is shown in Sect. 4.3.

If the physical event is bound-bound transition, bound-free transition, or free-free transition, then an unpolarized VP propagating toward the viewing direction is created. The co-moving frame frequency of the VP is the same as the OP after the physical event, the energy of the VP is

$$\bar{E}_{VP} = \frac{1}{4\pi}\bar{E}_{OP}. \tag{4.5}$$

[1] In the following, co-moving frame quantities are denoted with a bar, whereas non-bar quantities refer to the rest frame.

After creation, the VPs propagate to the viewing direction with the speed of light, and the VPs are not considered in the follow-up computation of the plasma level and ionization states. The total optical depth of a VP is integrated over the trajectory:

$$\tau_{tot} = \sum_j d_j \alpha_{tot,j} , \tag{4.6}$$

where d_j is the length of the j-th line segment of the VP trajectory, $\alpha_{tot,j}$ is the total opacity at the j-th line segment, and the rest frame opacity is calculated from the co-moving frame opacity:

$$\alpha_{tot,j}(\nu) = (\bar{\nu}/\nu)\bar{\alpha}_{tot,j}(\bar{\nu}) . \tag{4.7}$$

The line segment is split when:

- The VP reaches the boundary of a volume cell.
- The VP reaches the end of a time step.
- The co-moving frame frequency of the VP reaches the boundary of the frequency grid of α_{tot}.

The final spectrum is the sum of the escaped VPs:

$$\begin{pmatrix} \mathcal{I} \\ \mathcal{Q} \\ \mathcal{U} \\ \mathcal{V} \end{pmatrix} = \frac{1}{4\pi r^2 \Delta t \Delta \nu} \sum_i E_{VP,i} s_{VP,i} e^{-\tau_{tot,i}} , \tag{4.8}$$

and the summation is over all the VPs in the frequency bin $[\nu - \Delta\nu/2, \nu + \Delta\nu/2)$, and arrival time interval $[t - \Delta t/2, t + \Delta t/2)$.

We implement the EBT into SEDONA. Comparing to the EBT recipe used in ARTIS [6], we did not include two features:

- The VP is deleted when its τ_{tot} reaches 10.
- The VP is not created if its rest frame wavelength will not be in the 3500–10,000 Årange.

According to [6], these two features could increase the computation speed by a factor of ~4, and the $\tau_{tot} < 10$ criterion does not affect the accuracy in most of the cases.

4.2 The Integral-Based Technique

The proposed IBT uses integrated packets (IPs), which is an upgraded version of VP, to compute spectra at specified viewing directions with significantly higher efficiency than VPs. Details are illustrated in the following subsections.

4.2.1 *The Integrated Packets*

The IP computation is based on a universal frequency grid $[\nu_0, \nu_1 \ldots \nu_N]$. The frequency ratio between the adjacent pixels is a constant C, satisfying:

$$\nu_{n+1}/\nu_n = C. \tag{4.9}$$

Using this frequency grid, the Doppler shift of the variables could be simplified to the shifting of the pixels and saves computation time.

The radiation energy in the rest frame (denoted as $\boldsymbol{R}(\vec{x}, t, \nu)$) from a volume cell at coordinate $\vec{x}$, time t, and frequency ν toward the viewer consists of isotropic unpolarized radiation and scattering radiation:

$$\boldsymbol{R}(\vec{x}, t, \nu) = \Delta v \Delta t \left(\frac{\bar{\nu}}{\nu}\right)^{-2} \left(\bar{j}_{emi}(\vec{x}, t, \bar{\nu}) \begin{pmatrix} 1 \\ 0 \\ 0 \\ 0 \end{pmatrix} + \alpha_T \int_\Omega \bar{\boldsymbol{I}}(\vec{x}, t, \bar{\theta}_{in}, \bar{\varphi}_{in}, \bar{\nu}) \boldsymbol{P}(\bar{\theta}_{in}, \bar{\varphi}_{in}, \bar{\theta}, \bar{\varphi}) d\Omega \right), \tag{4.10}$$

where $\bar{\boldsymbol{I}}$ is the specific intensity as a four vector of the Stokes parameters; the $\bar{j}_{emi}$ function is the unpolarized isotropic emission term, which includes all the emission from bound-bound, bound-free, and free-free transitions; the $\boldsymbol{P}(\bar{\theta}_{in}, \bar{\varphi}_{in}, \bar{\theta}, \bar{\varphi})$ function is a combination of the Rayleigh scattering phase matrix and rotation matrix discussed in Sect. 4.3; α_T is the Thomson scattering opacity; the integral is over all the incident directions $(\bar{\theta}_{in}, \bar{\varphi}_{in})$; the $(\bar{\nu}/\nu)^{-2}$ term is the correction of the Doppler effect (e.g., eq. (1–3) [7]; see also p. 31,33,495–496 [8]).

4.2.2 *Isotropic Emission*

The isotropic emission term could be calculated by the summation of the VPs from the same physical events:

$$\Delta v \Delta t \left(\frac{\bar{\nu}}{\nu}\right)^{-2} \bar{j}_{emi}(\vec{x}, t, \bar{\nu}) = \left(\frac{\bar{\nu}}{\nu}\right)^{-2} \sum_i \frac{\bar{E}_{VP,i}}{\Delta \bar{\nu}}, \tag{4.11}$$

where the summation is over all the VPs within the frequency bin $[\bar{\nu} - \Delta\bar{\nu}/2, \bar{\nu} + \Delta\bar{\nu}/2)$, the volume cell Δv, and the time step bin $[t - \Delta t/2, t + \Delta t/2)$. Note that the VPs created in this step are not propagated in the following calculations, and the computation resources are recycled. This recipe is suitable to complicated physical events (i.e., macroatom [9]).

In SEDONA calculation, we adopt the line expansion opacity (eq. (8) [10]) for the bound-bound transition, and the absorbed photons due to bound-free and free-free transition are treated to reemit as thermalized photon. Therefore, the isotropic emission term is calculated from the current plasma state with $\bar{j}_{emi}(\vec{x}, t, \bar{\nu}) = B(\bar{\nu})\bar{\alpha}_{abs}(\vec{x}, t, \bar{\nu})$, where $B(\bar{\nu})$ is the black body radiation spectrum and $\bar{\alpha}_{abs}$ is the total absorptive opacity including bound-bound, bound-free, and free-free transitions. (Note that there is no Thomson scattering.) This recipe could generate emission term without Monte Carlo noise and could accelerate the program by avoiding the time-consuming summation computation.

4.2.3 The Path Integral

The radiation energy from Thomson scattering is calculated from the path integral of the OPs:

$$\begin{aligned}\Delta v \Delta t_s \left(\frac{\bar{\nu}}{\nu}\right)^{-2} \sigma_T \int_{\Omega} \bar{\boldsymbol{I}}(\vec{x}, t, \bar{\theta}_{in}, \bar{\varphi}_{in}, \bar{\nu}) \boldsymbol{P}(\bar{\theta}_{in}, \bar{\varphi}_{in}, \bar{\theta}, \bar{\varphi}) d\Omega \\ = \frac{1}{4\pi \Delta \bar{\nu}} \sum_{i,j} \bar{E}_i \bar{d}_{i,j} \bar{\alpha}_T \bar{\boldsymbol{s}}_{i,j} \boldsymbol{P}(\bar{\theta}_{i,j}, \bar{\varphi}_{i,j}, \bar{\theta}, \bar{\varphi}),\end{aligned} \tag{4.12}$$

where $d_{i,j}$ is the rest frame length of the j-th line segment of the i-th OP trajectory in the volume cell, and in the time step, and the summation is over all the trajectories within the frequency bin $[\bar{\nu} - \Delta\bar{\nu}/2, \bar{\nu} + \Delta\bar{\nu}/2)$, the volume cell Δv, the time step bin $[t_s, t_s + \Delta t_s)$.

For each volume cell and at each time step, an IP is created. The initial coordinate of the IP is randomly chosen in the volume cell, and the initial time of the IP is randomly chosen in the time step. The radiation energy $\boldsymbol{R}$ over the frequency grid is also stored in each IP.

4.2.4 The Propagation of Integral Packets

The IPs propagate toward the viewer with the speed of light, and the radiation energy $\boldsymbol{R}$ is reduced over the trajectory. Similar to the propagation of the VPs, the trajectory of an IP is split into line segments by the edges of time steps and volume cells. Moreover, the trajectory is further split to satisfy the criterion:

$$\left|\frac{\bar{\nu}_{begin} - \bar{\nu}_{end}}{\bar{\nu}_{end}}\right| \leq C - 1, \tag{4.13}$$

where $\bar{\nu}_{begin}$ is the co-moving frame frequency at the beginning of the line segment, $\bar{\nu}_{end}$ is the co-moving frame frequency of the same rest frame frequency ν at the ending of the line segment, and the constant C is defined in Eq. 4.9.

In the calculation of VP in Eqs. 4.6 and 4.8, the total opacity is accumulated over the trajectory, which computationally requires one more double precision floating point number per VP to store the opacity. This process reduces the computation time on exponential at the cost of memory space. While in the calculation of IP, the radiation energy is updated at every line segment on the trajectory without accumulating the total opacity over the trajectory:

$$\boldsymbol{R}_{new} = \boldsymbol{R}_{old} e^{-d_j \alpha_{tot,j}} , \tag{4.14}$$

where d_j is the length of the current j-th line segment over the trajectory, and $\alpha_{tot,j}$ is the total opacity at the j-th line segment. The final spectrum is the sum of the escaped IPs:

$$\begin{pmatrix} \mathcal{I} \\ \mathcal{Q} \\ \mathcal{U} \\ \mathcal{V} \end{pmatrix} = \frac{1}{4\pi r^2 \Delta t \Delta \nu} \sum_i \boldsymbol{R}_i , \tag{4.15}$$

and the summation is over all the IPs in the arrival time interval $[t - \Delta t/2, t + \Delta t/2)$. Note that the grid size of arrival time Δt should be much larger than the time step size Δt_s to avoid grid mismatch issue, while this issue is not significant in DCT or EBT calculations.

4.3 Rayleigh Scattering

The Thomson scattering between free electron and photon is calculated by the Rayleigh scattering phase matrix:

$$\boldsymbol{P}_R(\Theta) = \frac{3}{4} \begin{pmatrix} cos^2\Theta + 1 & cos^2\Theta - 1 & 0 & 0 \\ cos^2\Theta - 1 & cos^2\Theta + 1 & 0 & 0 \\ 0 & 0 & 2cos\Theta & 0 \\ 0 & 0 & 0 & 2cos\Theta \end{pmatrix} , \tag{4.16}$$

where Θ is the angle between the incident light and the emergent light. Before the calculation of Rayleigh scattering phase matrix, the reference axis of the incident light should be rotated into the scattering plane, and the polarization state is changed by the rotation matrix:

$$\boldsymbol{L}(\phi_{in}) = \begin{pmatrix} 1 & 0 & 0 & 0 \\ 0 & cos2\phi_{in} & sin2\phi_{in} & 0 \\ 0 & -sin2\phi_{in} & cos2\phi_{in} & 0 \\ 0 & 0 & 0 & 1 \end{pmatrix} , \tag{4.17}$$

where ϕ_{in} is the rotation angle between the observer's reference axis to the scattering plane reference axis. A similar rotation matrix of $\boldsymbol{L}(\pi - \phi_{out})$ is also applied to the emergent light, in order to convert the polarization state into the observer's reference axis. Therefore, the full Rayleigh scattering formula in the SN radiative transfer simulation is written as

$$\bar{\boldsymbol{S}}_{out} = \boldsymbol{L}(\pi - \bar{\phi}_{out})\boldsymbol{P}_R(\bar{\Theta})\boldsymbol{L}(\bar{\phi}_{in})\bar{\boldsymbol{S}}_{in} . \tag{4.18}$$

Note that the Rayleigh scattering happens in the co-moving frame, and all the variables in the above equation are measured in the co-moving frame. In Eqs. 4.4, 4.10, and 4.12, the operator is defined as $\boldsymbol{P}(\bar{\theta}_{in}, \bar{\phi}_{in}, \bar{\theta}_{out}, \bar{\phi}_{out}) = \boldsymbol{L}(\pi - \bar{\phi}_{out})\boldsymbol{P}_R(\bar{\Theta})\boldsymbol{L}(\bar{\phi}_{in})$.

Replacing the axis rotation angles and scatter angle ($\bar{\phi}_{in}$, $\bar{\phi}_{out}$, $\bar{\Theta}$) with the propagating direction of the incident OP ($\bar{\theta}_{in}, \bar{\varphi}_{in}$) and the propagating direction of the emergent VP/IP ($\bar{\theta}, \bar{\varphi}$), the expression of the operator $\boldsymbol{P}(\bar{\theta}_{in}, \bar{\phi}_{in}, \bar{\theta}, \bar{\phi})$ becomes [1]

$$\begin{aligned}\bar{I}_{out} =& \frac{3}{8\pi}\Big(\left((l,l)^2 + (l,r)^2\right)\bar{I}_{in} + \left((r,l)^2 + (r,r)^2\right)\bar{Q}_{in} \\ &+ ((l,l)(r,l) + (l,r)(r,r))\,\bar{U}_{in}\Big)\end{aligned} \tag{4.19}$$

$$\begin{aligned}\bar{Q}_{out} =& \frac{3}{8\pi}\Big(\left((l,l)^2 - (l,r)^2\right)\bar{I}_{in} \\ &+ \left((r,l)^2 - (r,r)^2\right)\bar{Q}_{in} + ((l,l)(r,l) - (l,r)(r,r))\,\bar{U}_{in}\Big)\end{aligned} \tag{4.20}$$

$$\bar{U}_{out} = \frac{3}{8\pi}\left(2(l,l)(l,r)\bar{I}_{in} + 2(r,r)(r,l)\bar{Q}_{in} + ((l,l)(r,r) + (r,l)(l,r))\,\bar{U}_{in}\right) \tag{4.21}$$

$$\bar{V}_{out} = \frac{3}{8\pi}\left((l,l)(r,r) - (r,l)(l,r)\right)\bar{V}_{in} , \tag{4.22}$$

where (l,l), (r,l), (l,r), (r,r) are

$$(l,l) = sin(\bar{\theta})sin(\bar{\theta_{in}}) + cos(\bar{\theta})cos(\bar{\theta_{in}})cos(\bar{\varphi_{in}} - \bar{\varphi}) \tag{4.23}$$

$$(r,l) = cos(\bar{\theta})sin(\bar{\varphi_{in}} - \bar{\varphi}) \tag{4.24}$$

$$(l,r) = -cos(\bar{\theta_{in}})sin(\bar{\varphi_{in}} - \bar{\varphi}) \tag{4.25}$$

$$(r,r) = cos(\bar{\varphi_{in}} - \bar{\varphi}) . \tag{4.26}$$

4.4 Model Validation

In this section, we prepare two toy models to validate the IBT computation results. The first model is type Ia supernova delayed-detonation model DDC10. DDC10 is calculated from hydrodynamic simulations in [11] and adopted as a benchmark in [12] to compare the difference between radiative transfer programs. The model is expanded into 3-D space of 60 × 60 × 60 Cartesian grid under spherical symmetry, and the boundary velocity is limited to 25,000 km/s. Figure 4.1 shows the abundance of several elements and the density profile of the DDC10 model. The second model is a modification based on the DDC10 model, and the y and z coordinates are re-projected with $y_{new} = 1.3y$ and $z_{new} = z/1.3$, in order to introduce 3-D features. This model is also in a 60 × 60 × 60 Cartesian grid, and the boundary velocity is 32,500 km/s. The model geometry is shown in the right panel of Fig. 4.1.

In the following text, the spherical symmetric DDC10 model is denoted as D1D, and the 3-D modified model is denoted as D3D. Both the models undergo 3-D time-dependent radiative transfer calculations in SEDONA starting from 0.976 days after the explosion. Before 6.6 days, the simulation time step is a logarithmic with $\Delta t_s/t_s = 0.03$. After 6.6 days, the simulation time step is $\Delta t_s = 0.2$ days. The arrival time bin size is $\Delta t = 1$ day, which satisfies $\Delta t \geq 5\Delta t_s$. The frequency grid is from 1×10^{14} Hz to 5×10^{15} Hz (30,000 Åto 600 Å) with $\Delta\nu/\nu = 0.002$.

4.4.1 *Spherical Symmetry*

Because the linear polarization components are canceled out, a spherical symmetric model should produce zero polarization. Using IBT, we calculate the spectropolarimetry of the D1D model to test if this algorithm could recover zero polarization, each time step generates 4×10^5 gamma-ray energy packets. Figure 4.2 shows the spectra and polarization percentage at different times after the explosion. We notice the polarization signal at most the arrival time, and wavelengths are consistent to the

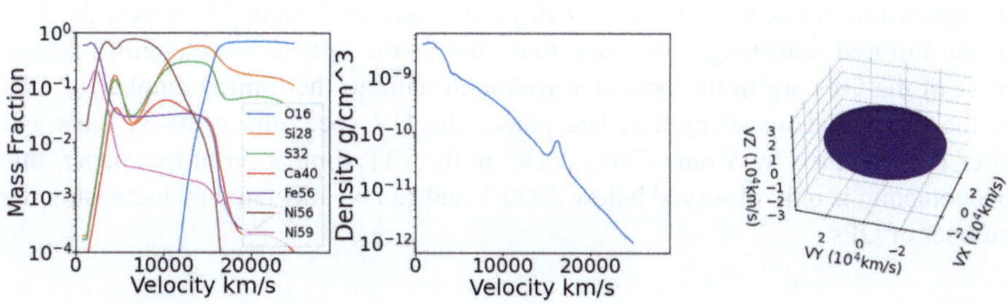

Fig. 4.1 Left panel: the abundances of several isotopes in the DDC10 model. Middle panel: the density structure of the DDC10 model. Right panel: the geometry of the modified DDC10 model with 3-D structures

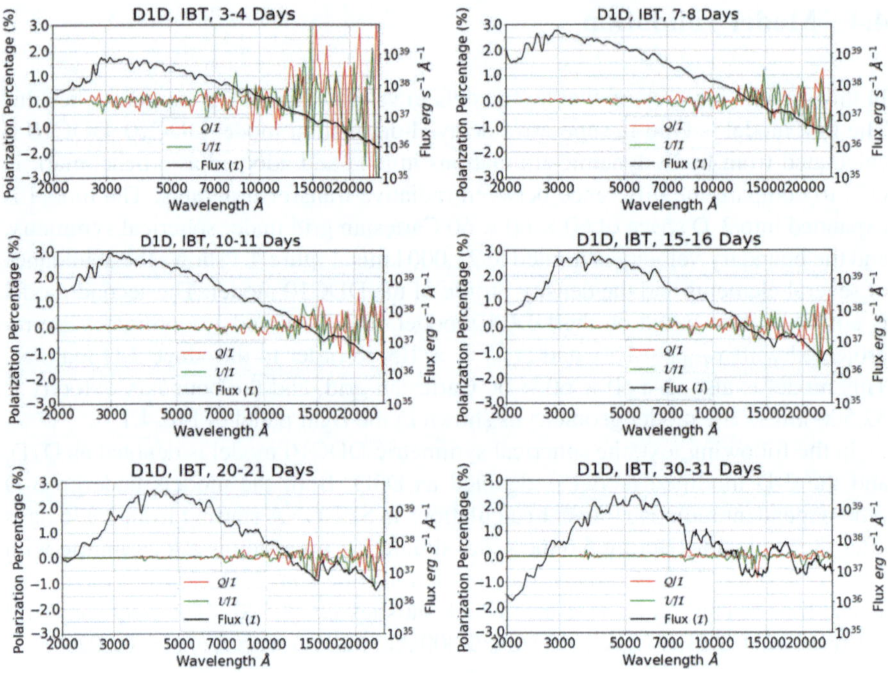

Fig. 4.2 The spectral flux ($\mathcal{I}$) and the polarization percentage ($\mathcal{Q}/\mathcal{I}$, $\mathcal{U}/\mathcal{I}$) of the spherical symmetric D1D model at different times. The spectra are calculated by IBT with 6×10^6 OPs per simulation time step. The arrival time bin relative to the explosion time is labeled in the title of each subplots

theoretical zero polarization result. Moreover, the simulation error between 7 days and 31 days and between 2000 and 10,000 Å wavelength range, at which most of the SNe Ia spectropolarimetry observations are made [13], is as low as ~0.2%. Notably, the spectra in the ultraviolet wavelength between 7 days and 21 days also show high S/N. At early phase of SNe Ia, most of the OPs are generated below the photosphere, and therefore the radiation energy $\boldsymbol{R}$ above the photosphere has low S/N. As a result, the spectrum shows low S/N at 3–4 days after the explosion. Moreover, the S/N in the infrared wavelength is lower than that in the optical wavelength, because most of the OPs are in the optical wavelength. Due to the limited number of OPs in the ultraviolet wavelength at late phase, the UV spectrum at 30–31 days and after is dominated by Monte Carlo noise. In the IBT simulation of this paper, this phenomenon is only observed below 2000 Å and can be alleviated by increasing the number of OPs.

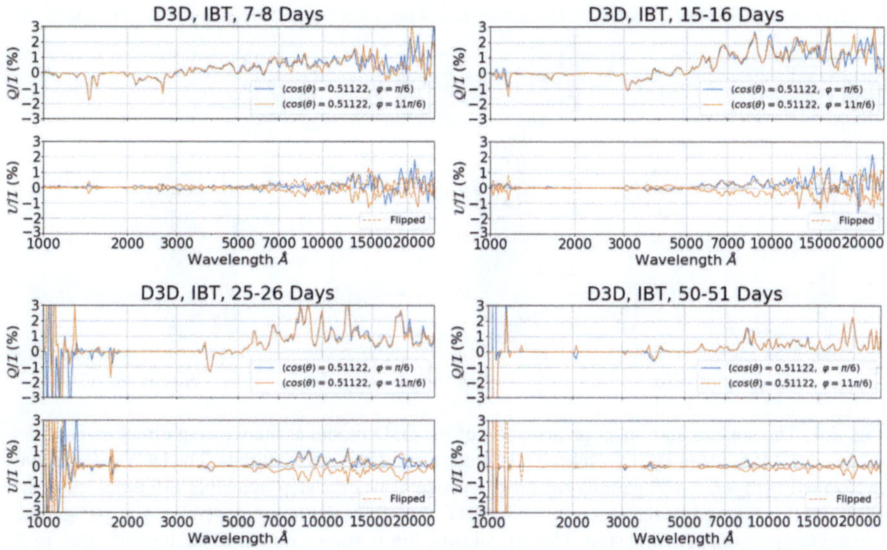

Fig. 4.3 The spectropolarimetry of the D3D model at the two symmetric viewing directions $(cos(\theta) = 0.51122,\ \varphi = \pi/6)$ and $(cos(\theta) = 0.51122,\ \varphi = 11\pi/6)$. The arrival time bin at each panel is labeled in the title of figures. In the subfigures of Stokes $\mathcal{U}/\mathcal{I}$ terms, the flipped values at one of the viewer points are shown as dashed line, in order to make direction comparisons. The polarization percentages of $\mathcal{Q}/\mathcal{I}$ and $\mathcal{U}/\mathcal{I}$ are averaged using a 7-pixel bin

4.4.2 Mirror Symmetry

In this section, we use IBT to calculate the spectropolarimetry at two mirror symmetric viewing directions of the D3D model. Based on the mirror symmetry of the D3D model on the X-Z, Y-Z, and X-Y planes, the spectropolarimetry at the viewing directions $(cos(\theta) = 0.51122,\ \varphi = \pi/6)$ and $(cos(\theta) = 0.51122,\ \varphi = 11\pi/6)$ should have the same Stokes $\mathcal{Q}$ terms and inverted $\mathcal{U}$ terms. Figure 4.3 shows the spectropolarimetry at the two symmetric viewing directions. We notice the simulation results are consistent to the theoretical predictions of the symmetry at above 2000 Å wavelength from early phase (7–8 days after the explosion) to late phase (50–51 days after explosion). Similar to the spherical symmetric validation results in Fig. 4.2, the ultraviolet spectropolarimetry below 2000 Å after ~25 days is dominated by Monte Carlo noise.

4.4.3 Cross Comparison

In this section, we compare the simulation results from EBT, IBT, and DCT methods using the D3D model. Figure 4.4 shows the spectra and linear polarization time sequence calculated by the three methods. In the DCT calculation, the viewing

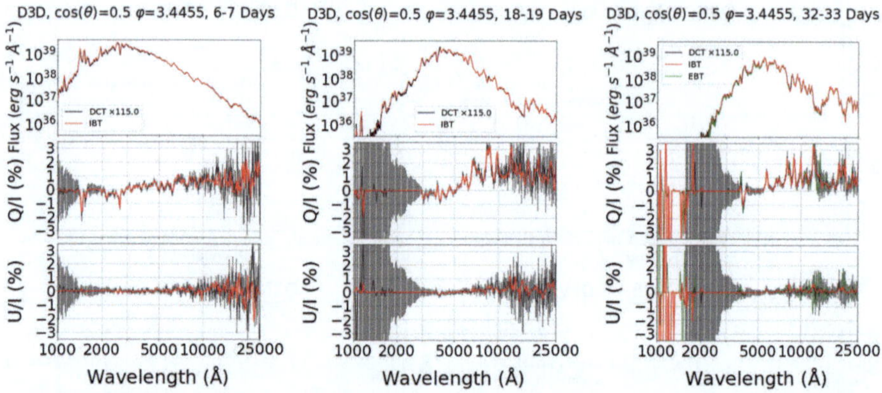

Fig. 4.4 The spectra and linear polarization of the D3D model at the viewing direction $cos(\theta) = 0.5$, $\varphi = 3.4455$, using DCT (black), EBT (green), and IBT (red) methods. The DCT calculation is performed with 4.6×10^8 gamma-ray energy packets per time step (which is 115 times more than EBT or IBT, marked in the legend), while EBT and IBT calculation generates 4×10^6 gamma-ray energy packets per time step. The arrival time bin is shown on top of each panel. Due to the memory limit, EBT calculation is only started at 25 days after the explosion and the EBT results are not shown at 6–7 days and 18–19 days. The polarization percentage is binned with seven pixels

direction is split into a 10×10 bin, and generates 4.6×10^8 gamma-ray energy packets per time step to guarantee the signal-to-noise ratio (S/N), while EBT and IBT only generate 4×10^6 gamma-ray energy packets per time step. We notice IBT could reproduce the spectra and Stokes parameter Q and U terms with high S/N comparable to DCT with ~115 times more OPs from 6 days to 33 days after the explosion and from ultraviolet to infrared wavelength. After ~32 days, the three methods are failed between 1000 and 2000 Ådue to the limit of the OPs.

Due to the high opacity at the early phase of SN, the OPs undergo multiple scattering events near the core of SN, which result in the uncontrolled number of VPs in EBT calculation and cause memory usage issues. Therefore, this example EBT calculation starts at 25 days after the explosion due to the limit of hardware.

4.5 Computation Performance

4.5.1 Computation Speed Comparison

The computation speed of the three methods is measured using one 48-core compute node of Grace supercomputer in TAMU HPRC.[2] Each node has two sockets of Intel Xeon 6248R CPUs, in total 48 cores, and has 384 GBs of RAM memory.

[2] https://hprc.tamu.edu/

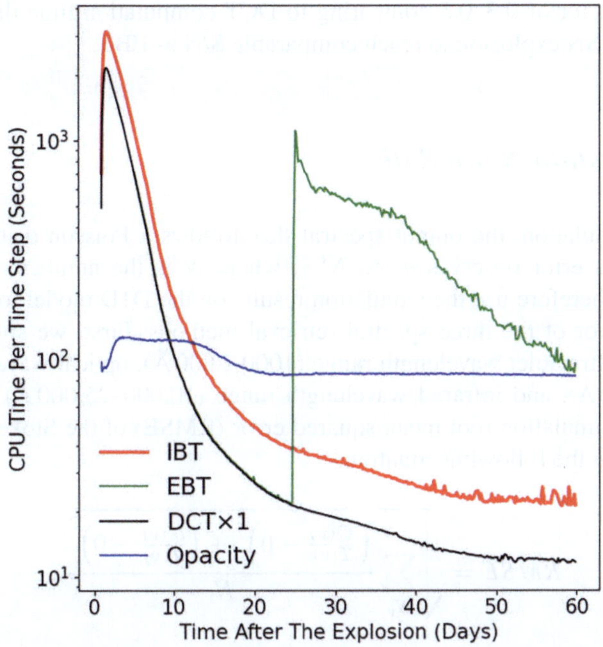

Fig. 4.5 Comparison of particle propagation computation time at different time steps using the DCT (black), EBT (green), and IBT (red) methods. Each time step generates 4×10^6 gamma-ray energy packets in the three simulations. The computation time of the plasma state and opacity is also shown in the figure with ultramarine color. Note that the EBT calculation is not activated until 25 days after the explosion, and the green curve before day 25 is overlapped to the black curve

The majority of the computation time in a 3-D time-dependent radiative transfer computation consists of two components: (1) the computation of plasma state, level population, and opacity of the grid; (2) the propagation of energy packets including gamma-ray packets, OPs, VPs, or IPs. The choice of the spectral retrieval methods only affects the propagation time of energy packets.

Figure 4.5 shows the computation time of particle propagation at each time step as a function of the time after explosion using the EBT and IBT methods as calculated in Sect. 4.4.3, and the corresponding DCT computation time with 4×10^6 gamma-ray energy packets per time step, the same number of packets as the EBT and IBT results. The computation time of the plasma state and opacity does not change drastically at different time steps. In comparison, the computation time of the particle propagation increases at the first $\sim$3 days, and because the opacity is high and OPs are unable to escape the SN ejecta, the physical events associated to these accumulated OPs lead to an increased computation time. With the expansion of the SN ejecta, the opacity decreases and the accumulated OPs escape the SN ejecta, and the computation time decreases. Notably, the computation time of EBT is a factor of $\sim$30 higher than DCT around 25 days after the explosion and is a factor of $\sim$10 higher than DCT around 60 days. In contrast, the computation time of IBT is a factor of 0.3–0.5 comparing to DCT computation time throughout the time after the SN explosion to reach comparable S/N as EBT.

4.5.2 Signal-to-Noise Ratio

In MCRT simulation, the output spectral flux follows a Poisson distribution, and the simulation error observes $\sigma \propto N^{0.5}$, where N is the number of the energy packets. We therefore use the simulation results on the D1D model to measure the simulation error of the three spectral retrieval methods. First, we split the output spectra into ultraviolet wavelength range (1000–2000 Å), optical wavelength range (2000–10,000 Å), and infrared wavelength range (10,000–25,000 Å). Second, we measure the simulation root mean squared error (RMSE) of the Stokes parameters $\mathcal{Q}$ and $\mathcal{U}$ using the following equation:

$$RMSE = \sqrt{\sum_{N_\nu} \frac{\left(\frac{\mathcal{Q}(\nu)}{\mathcal{I}(\nu)} - 0\right)^2 + \left(\frac{\mathcal{U}(\nu)}{\mathcal{I}(\nu)} - 0\right)^2}{N_\nu}}, \tag{4.27}$$

where N_ν is the number of pixels in the selected wavelength range, and the summation is over all the pixels in the selected wavelength range. Note that the theoretical solution of the D1D model is no polarization, which results in the zero term in the equation. The RMSE observes the inverse square root relation with the number of packets, $RMSE \propto N^{-0.5}$.

Figure 4.6 shows the relation between the number of gamma-ray energy packets per time step and the resulting RMSE at 10–11 days after the explosion using DCT or IBT, and at 37–38 days after the explosion using DCT, EBT, or IBT. We notice the result from IBT could reproduce the $RMSE \propto N^{-0.5}$ relation in most of the time and wavelength ranges. The result from DCT also reproduces the $RMSE \propto N^{-0.5}$ relation in optical and infrared wavelength ranges when the number of packets is large enough. However, in the ultraviolet wavelength, and in the optical wavelength with the number of packets per step smaller than $\sim 4 \times 10^5$, the inverse square root relation is not observed due to the large simulation error. The result from EBT also reproduces the $RMSE \propto N^{-0.5}$ relation in optical and infrared wavelength ranges at 37–38 days after the explosion. In the ultraviolet wavelength range at 37–38 days after the explosion, none of the methods could simulate good S/N spectra to reproduce the $RMSE \propto N^{-0.5}$ relation.

A direct comparison of the computation efficiency between the three methods could be made using the RMSE values measured in the optical and infrared wavelength ranges. The RMSE of IBT is smaller than the RMSE of DCT with a factor of 20−30, resulting in a factor of ~400 less packets to calculate spectral time sequences with similar S/N. The RMSE of EBT is smaller than the RMSE of DCT with a factor of ~10, resulting in a factor of ~100 less packets to calculate spectral

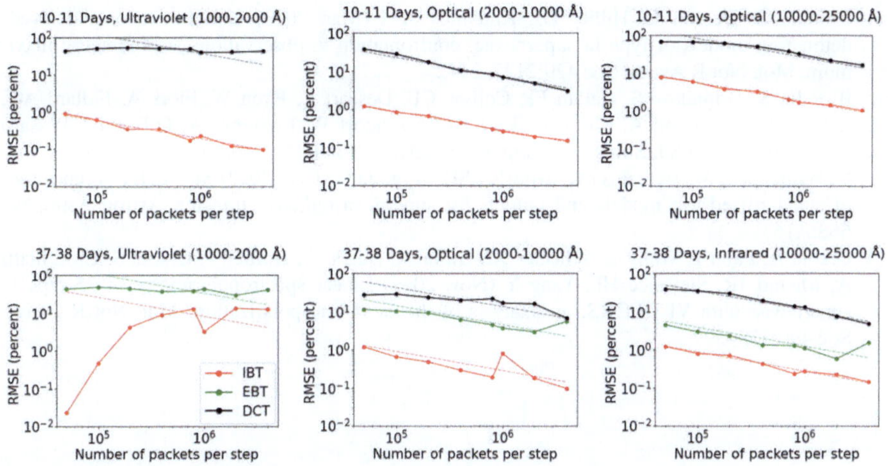

Fig. 4.6 The RMSE of DCT, EBT, and IBT simulation results on the D1D model, as a function of the number of gamma-ray energy packets generated per time step. The polarization spectra are binned with seven pixels. Upper panel shows the RMSE in the time bin of 10–11 days after the explosion, and lower panel shows the RMSE in the time bin of 37–38 days after the explosion. From left to right, the wavelength range is ultraviolet, optical, and infrared. The dashed line shows the theoretical inverse square root relation between RMSE and the number of packets ($RMSE \propto N^{-0.5}$) for comparison

time sequences with similar S/N. The accuracy measurement is consistent to the reports in Table 1 of [6].

References

1. Chandrasekhar S (1960) Radiative transfer. Dover Publications. ISBN 9780486605906
2. Lucy LB (May 1999) Improved Monte Carlo techniques for the spectral synthesis of supernovae. Astron Astrophys 345:211–220
3. Sim SA, Miller L, Long KS, Turner TJ, Reeves JN (May 2010) Multidimensional modelling of X-ray spectra for AGN accretion disc outflows - II. Mon Not R Astron Soc 404:1369–1384
4. Long KS, Knigge C (Nov. 2002) Modeling the spectral signatures of accretion disk winds: a new Monte Carlo approach. Astrophys J 579:725–740
5. Kerzendorf WE, Sim SA (2014) A spectral synthesis code for rapid modelling of supernovae. Mon Not R Astron Soc 440:387–404
6. Bulla M, Sim SA, Kromer M (June 2015) Polarization spectral synthesis for Type Ia supernova explosion models. Mon Not R Astron Soc 450:967–981
7. Castor JI (Dec. 1972) Radiative transfer in spherically symmetric flows. Astrophys J 178:779–792
8. Mihalas D (1978) Stellar atmospheres. W.H. Freeman and Company, San Francisco
9. Lucy LB (Mar. 2002) Monte Carlo transition probabilities. Astron Astrophys 384:725–735
10. Kasen D, Thomas RC, Nugent P (Nov. 2006) Time-dependent Monte Carlo radiative transfer calculations for three-dimensional supernova spectra, light curves, and polarization. Astrophys J 651:366–380

11. Blondin S, Dessart L, Hillier DJ, Khokhlov AM (Mar. 2013) One-dimensional delayed-detonation models of type Ia supernovae: confrontation to observations at bolometric maximum. Mon Not R Astron Soc 429:2127–2142
12. Blondin S, Blinnikov S, Callan FP, Collins CE, Dessart L, Even W, Flörs A, Fullard AG, Hillier DJ, Jerkstrand A, Kasen D, Katz B, Kerzendorf W, Kozyreva A, O'Brien J, Pássaro EA, Roth N, Shen KJ, Shingles L, Sim SA, Singhal J, Smith IG, Sorokina E, Utrobin VP, Vogl C, Williamson M, Wollaeger R, Woosley SE, Wygoda N (Dec. 2022) StaNdaRT: a repository of standardised test models and outputs for supernova radiative transfer. Astron Astrophys 668:A163
13. Cikota A, Patat F, Wang L, Wheeler JC, Bulla M, Baade D, Höflich P, Cikota S, Clocchiatti A, Maund JR, Stevance HF, Yang Y (Nov. 2019) Linear spectropolarimetry of 35 type Ia supernovae with VLT/FORS: an analysis of the Si II line polarization. Mon Not R Astron Soc 490:578–599

Chapter 5
Constraining Type Ia Supernova Delay Time with Spatially Resolved Star Formation Histories

Abstract We present the delay time distribution (DTD) estimates of Type Ia supernovae (SNe Ia) using spatially resolved SN Ia host galaxy spectra from MUSE and MaNGA. By employing a grouping algorithm based on k-means and earth mover's distances (EMD), we separated the host galaxy stellar population age distributions (SPADs) into spatially distinct regions and used maximum likelihood method to constrain the DTD of SN Ia progenitors. When a power-law model of the form $DTD(t) \propto t^s (t > \tau)$ is used, we found an SN rate decay slope $s = -1.41^{+0.32}_{-0.33}$ and a delay time $\tau = 120^{+142}_{-83} Myr$. Moreover, we tested other DTD models such as a broken power-law model and a two-component power-law model and found no statistically significant support to these alternative models.

5.1 Data Reduction

5.1.1 Sample Selection

Most of the data used in our research were taken by the multiunit spectrograph explorer (MUSE), an integral-field unit (IFU) facility mounted on the 8.4 meter telescope Yepun (UT4) at very large telescopes (VLT). We used all publicly available SNe Ia listed on Transient Name Server (TNS),[1] and cross-matched them with available MUSE data. For the matched SN-host pairs, the following selection criteria were applied:

- The time difference between SN discovery and IFU observation is at least 100 days to avoid SN light from contaminating the host galaxy's spectra.

[1] https://wis-tns.weizmann.ac.il

X. Chen, *Deep Learning on Type Ia Supernovae*, Springer Theses,
https://doi.org/10.1007/978-3-032-13032-7_5

- The spectral type of the SN is normal SN Ia according to both TNS[1] and SIMBAD.[2]
- More than 70% of the flux of the host galaxy is covered by the MUSE field of view.
- After binning the spectral data cube by 3×3 spatial pixels, the signal-to-noise ratio (SNR) at the position of the SN is at least 0.8 per Å.
- The host galaxy has at least 200 spatial pixels with $SNR > 0.8$ per Å in the data spatially binned by 3×3.

We found 100 SNe Ia in 96 host galaxies (four galaxies host 2 SNe in each) satisfying the above criteria, and they were observed by 37 VLT observation programs. The list of all SNe and host galaxies are shown in Appendix A Table A.1. Among all these observations, four of them (SN 2011iv, SN 2011is, SN 2009ev, SN 1992A) employed the extended wavelength coverage from 465 nm to 930 nm using the `WFM-NOAO-E` instrumentation mode, one (SN 2000do) employed the adaptive optics (AO) technique using the `WFM-AO-N` instrumentation mode, and the rest were observed in natural seeing with a nominal wavelength range from 480 nm to 930 nm using the `WFM-NOAO-N` instrumentation mode. The spectral resolving power for both extended wavelength coverage and nominal coverage varies with wavelength, with resolving power $R = 2000\text{–}4000$, and all the observations used the wide field mode (`WFM`) with 1 square arcminute field of view (FoV). In Fig. 5.1, we show the histogram of the seeing distribution.

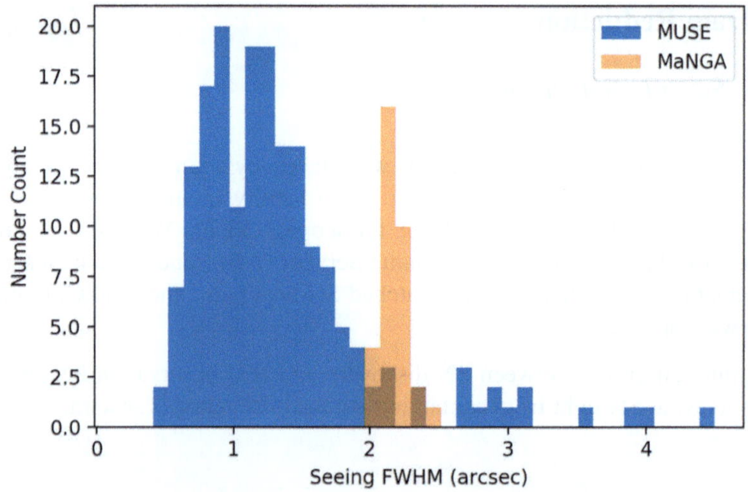

Fig. 5.1 The histogram of the seeings of the data used in our research. Blue histogram is for the MUSE dataset, and the orange histogram is for the MaNGA dataset

[2] http://simbad.u-strasbg.fr/simbad/

We downloaded the data product, which was reduced by the MUSE data reduction pipeline [1], from ESO Science Archive.[3] During the multiyear observation period, the version of the data reduction pipeline had changed from `v1.4` to `v2.8`, and we did not attempt to reduce the data again with the latest version of the pipeline. Moreover, some data cubes are stacks of multiple observations, which makes the spatial coverage of some data cubes larger than the FoV of the instruments.

In addition to the MUSE data, the mapping nearby galaxies at APO (MaNGA) [2] is a spectroscopic survey of galaxies using the 2.5 m telescope at Apache Point Observatory (APO) which also provides observations of SN-host galaxies. The survey has a hexagonal FoV with a diameter varying from 12 arcseconds to 32 arcseconds covering the wavelength range 3600 – 10300 Å with a resolving power $R \sim 2000$ and a spatial pixel scale of 0.5 arcseconds. We applied the same selection criteria (except binning of the spectral data cube) as used for the MUSE dataset and found 30 SN Ia host galaxies among 4824 galaxies in SDSS-DR15 [3], and the information on the SNe and host galaxies is tabulated in Appendix A Table A.2. The fully reduced data cubes, downloaded via `marvin` [4], were used to calculate the spatially resolved SPADs.

5.1.2 Spatially Resolved SPAD Calculation

As mentioned above, the original MUSE data cubes were binned by 3×3 spatial pixels to increase the SNR of each spectral element. The software Source Extractor [5] is used to identify the sky background close to the host galaxy and filter out the foreground stars. The pixels with spectral SNR above 0.8 per Å of the SN-host galaxies were used to calculate the SPADs. Note that some recent studies (e.g., [6]) on the stellar population with MUSE data employed Voronoi tessellation to build adaptive grids to spatially bin the spectra with lower SNRs. We did not adopt this method because it will mutilate spatial information which is important in our research. The MaNGA data cubes were not binned, and we used the mask provided in `marvin` to remove foreground stars and bad pixels. In all of the MaNGA survey targets, the galaxy covers the entire or most part of the FoV, and the foreground stars were identified in `marvin` program, so we did not use Source Extractor on these galaxies.

Penalized Pixel Fitting (`ppxf`) [7], which solves for a linear combination of simple stellar population (SSP) models from a stellar spectral library to fit the observed spectra, is used in our research to calculate the host galaxies' SPADs. We used the `E-MILES` stellar spectral library and assume Padova stellar evolutionary tracks and isochrones [8, 9] initial mass function (IMF), and $[\alpha/\text{Fe}] = 0$ for all the galaxies. The model spectral library consists of 50 age grids ranging from 63.10 Myr to 17.78 Gyr and seven metallicity grids ranging from $[M/H] = -2.32$

[3] http://archive.eso.org/

to $[M/H] = 0.22$. We used a 10 order multiplicative polynomial and no additive polynomial to fit the spectral continuum in the spectral fitting process.

In `ppxf`, the smoothness of the SPAD is controlled by the regularization parameter `regul`. We adopted the method introduced in [10] and [11] to modify the `regul` parameter. For the SPAD calculation of one spectrum, we first calculated a series of fitting spectra with `regul` changing from 2^{10} to 2^{-8} as a geometric sequence with common ratio $1/2$. Fitting spectra with $regul = 0$ is also calculated as the unregularized case. Then, we multiplied the flux level and set the $\chi^2(regul = 0)/N$ statistic for `regul=0` model to be 1, where N is the number of pixels for the observed spectra. Finally, the `regul` value when $\chi^2(regul) - \chi^2(regul = 0)$ of the corresponding fit is equal or close to $\sqrt{2N}$ is chosen for the spectral fit, as suggested in [12]. Considering there are 96 data cubes and each data cube typically consists of ~3000 spectra of the SN-host galaxy, searching for the optimal `regul` parameter for each individual spectrum is computationally prohibitive. Accordingly, for each galaxy we only searched for `regul` using the spectrum with the highest SNR and then applied the `regul` value to the SPAD calculation of the entire data cube. Note that while the intrinsic SPADs may be stochastic and discrete, the smoothed SPADs from our method may not precisely be the true SPADs of the galaxy. Smoothing can reduce the degeneracy between age and metallicity from bin to bin and enable the differential comparison of SPADs within each galaxy.

5.2 Methodology

In this study, a galaxy is spatially separated into stellar groups based on the SPADs we had deduced. Accordingly, we applied theoretical DTD models to calculate the SN rates of the stellar groups. Subsequently, we used maximum likelihood method to derive the optimal parameters of the DTD models to maximize the SN rates of the SN-related groups and minimize the SN rates of the stellar groups unrelated to the SNe Ia for all of the selected galaxies. In Sect. 5.2.1, we describe the Earth mover's distance (EMD) and show how that can be used as a method to calculate the difference between two SPADs. In Sect. 5.2.2, we present our algorithm of spatially separating a galaxy into different groups. In Sect. 5.2.3, we present the four DTD models used in our research. In Sect. 5.2.4, we present the likelihood function we have used for the maximize likelihood estimation.

5.2.1 Earth Mover's Distance

Earth mover's distance (EMD), also known as Wasserstein distance, was developed to evaluate the similarity among distributions and has been widely used in image recognition [13] and deep learning [14]. By definition, EMD measures the minimum amount of work required to change one distribution into the other.

For $P = \{(x_1, p_1), \ldots, (x_i, p_i), \ldots, (x_m, p_m)\}$ and $Q = \{(y_1, q_1), \ldots, (y_j, q_j), \ldots, (y_n, q_n)\}$ as two distributions, where (x_i, y_j) are the centers of data groups (i, j) and (p_i, q_j) are the probabilities of the groups, we can define a flow $F = [f_{ij}]$ between x and y which represents moving the probability in x_i to y_j. The flow is a feasible flow between P and Q when

$$f_{ij} \geq 0,\ 1 \leq i \leq m, 1 \leq j \leq n \tag{5.1}$$

$$\sum_{j=1}^{m} f_{ij} \leq p_i,\ 1 \leq i \leq m \tag{5.2}$$

$$\sum_{i=1}^{n} f_{ij} \leq q_j,\ 1 \leq j \leq n \tag{5.3}$$

$$\sum_{i=1}^{m}\sum_{j=1}^{n} f_{ij} = min\left(\sum_{i=1}^{m} p_i, \sum_{j=1}^{n} q_j\right). \tag{5.4}$$

For a feasible flow $F(x, y)$, the work done by the flow in matching P and Q distribution is

$$W(F, P, Q) = \sum_{i=1}^{m}\sum_{j=1}^{n} f_{ij}d_{ij}, \tag{5.5}$$

where d_{ij} is the distance between x_i and y_j. When a flow is an optimal flow $F^*(P, Q) = [f^*_{ij}](P, Q)$, which minimizes $W(F, P, Q)$, the EMD is defined as the work normalized by the optimal flow:

$$EMD(P, Q) = \frac{\sum_{i=1}^{m}\sum_{j=1}^{n} f^*_{ij}d_{ij}}{\sum_{i=1}^{m}\sum_{j=1}^{n} f^*_{ij}}. \tag{5.6}$$

As mentioned in Sect. 5.1.2, the spectral isochrone [8] used in our research is evenly sampled in $log_{10}(t)$ space with 50 grids between 63.10 Myr and 17.78 Gyr, and therefore we use $log_{10}(t_i) - log_{10}(t_j)$ for the distance d_{ij}. Moreover, all the SPADs are normalized so that $\sum_{i=1}^{50} p_i = \sum_{j=1}^{50} q_j = 1$ and $EMD(P, Q) = \sum_{i=1}^{50}\sum_{j=1}^{50} f_{ij}d_{ij}$.

5.2.2 K-Means Clustering

In the conventional k-means clustering algorithm, the distances are measured in Euclidean space. Given a dataset and a target number (k) of groups, the k-means algorithm first generates k "initial" centroids and then assigns data points into k

Algorithm 1: k-means based on EMD

input : Grouping number K
input : SPAD data cube
initialize $\mathcal{I}(x, y)_0 \leftarrow Rand[1, 2, \ldots, K]$;
initialize $i \leftarrow 1$;
while $\exists\, \mathcal{I}(x, y)_{i-1} \neq \mathcal{I}(x, y)_i$ **do**
    if $\exists\, n_0 \in [1, 2, \ldots, K], \mathcal{I}(x_0, y_0)_i = n_0$ & $\forall (x, y) \neq (x_0, y_0), \mathcal{I}(x, y)_i \neq n_0$ **then**
        $K \leftarrow K - 1$;
    end
    $MSPAD_n \leftarrow$ stellar mass averaged SPAD for the n-th group, $n \in [1, 2, \ldots, K]$;
    for *SPAD(x,y) in SPAD data cube* **do**
        $D_n \leftarrow EMD\,(SPAD(x, y), MSPAD_n)$;
        $\mathcal{I}(x, y)_{i+1} \leftarrow arg\ min_n\ D_n$;
    end
    $i \leftarrow i + 1$
end

groups according to the nearest centroids in Euclidean metric. The means are then updated using the centroids of the groups, and data points are iteratively reassigned into new groups.

A combination of EMD and k-means or k-medians algorithm has been used in sparse signal recovery problem [15, 16]. In our algorithm to separate a galaxy into different groups, an SPAD profile in a spatial pixel serves as a data point similar to the original k-means clustering algorithm, the mass-averaged SPAD of all data points in a group serves as the centroid in the original k-means clustering algorithm, and the EMD between an SPAD profile in a spatial pixel and the mass-averaged SPAD replaces the Euclidean distance in the original k-means algorithm. The algorithm is shown in Algorithm 1. The objective of this algorithm is to derive a map $\mathcal{I}(x, y)$ which records the group assignment of the stellar population at the pixel (x, y), with the inputs being the SPAD data cube and the total number of groups K. We introduced an extra operation, a group with only one pixel will be eliminated, and the data will be merged into the closest group.

We chose NGC 1516, the host galaxy of SN 2018ezx, to test the algorithm. The SNR of the MUSE observations of NGC 1516 is high, and there are 8682 spatial pixels satisfying our spectral selection limit (Sect. 5.1). It is an interacting galaxy with both galaxies inside the FoV of MUSE, which can potentially have two or more groups of SPADs in the data cube. We chose the total number of groups to be 2, 5, and 10 to test the algorithm and changed the initial $\mathcal{I}(x, y)_0$ multiple times to test the robustness of the results. Typical computational costs are 24 seconds, 98 seconds, and 351 seconds for 2, 5 and 10 groups, respectively, using one core of Intel Xeon E5-2670 v2. Also, with different random seeds for the initial $\mathcal{I}(x, y)_0$ maps, the final results are not affected. This test verifies that the algorithm can produce stable results for different initial $\mathcal{I}(x, y)_0$.

In Fig. 5.2, we show the final group maps $\mathcal{I}(x, y)$ of NGC 1516 with group numbers $K = 2, 5, 10$. The mean ages of the stellar populations are color encoded in

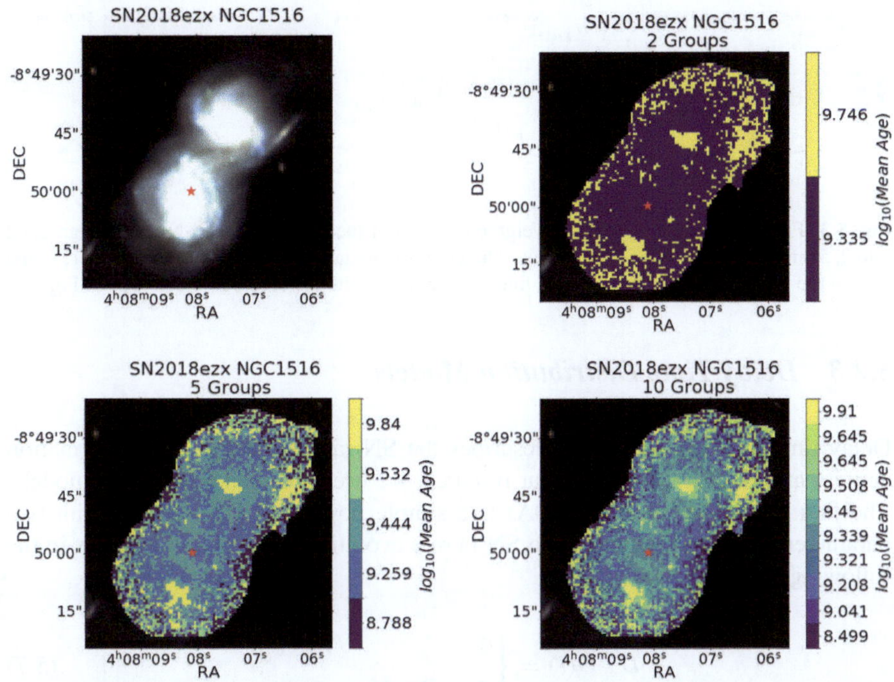

Fig. 5.2 Upper left: Image of NGC 1516, with the coordinate of SN 2018ezx marked by the red star. Upper right: The grouping map $\mathcal{I}(x, y)$ of two total groups for NGC 1516. Lower left: The grouping map $\mathcal{I}(x, y)$ of five total groups for NGC 1516. Lower right: The grouping map $\mathcal{I}(x, y)$ of 10 total groups for NGC 1516. The mean ages of the SPAD groups are encoded in different colors and are labeled in the color bars

which the ages increase going from purple to yellow. The center of the upper-right galaxy and the southern part of the lower-left galaxy are classified as the oldest stellar group in all cases. The outskirts of the galaxies show a mixture of young and old SPAD groups and could affect the group of the SN if the SN coordinate is in such a region. We surmise this phenomenon could due to the uncertainties in SPAD calculation introduced by the low SNR of the data. In Eq. 5.14 of Sect. 5.2.4, we will discuss the effect of observational seeing on the SN probability calculation to mitigate this problem.

We present the mass-averaged SPADs of each group in Fig. 5.3, and the ages are encoded in the colors of the curves. We notice that for the 2-group and 5-group separations, the averages of the SN-related groups peak at $10^{9.1}$ yr; for the separation with 10 groups, most of the pixels close to the SN coordinate belong to group 3 and the SFH of which also shows a peak at $10^{9.1}$ yr.

These exercises suggest that the ages of the SN progenitors can be estimated by comparing the SPADs of the host galaxies at the locations of the SNe with those away from them. However, not all the SNe-host galaxies show such a distinct signal, and maximum likelihood estimation is necessary to estimate the DTD.

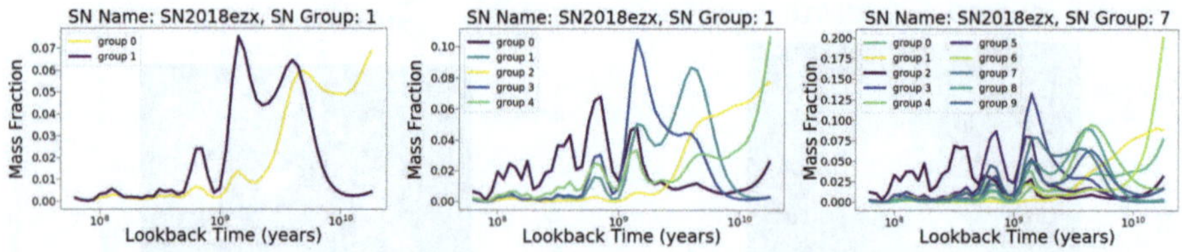

Fig. 5.3 From left to right, the mass-weighted SPADs of the host galaxy of SN 2018ezx grouped into 2, 5, and 10 total groups, respectively. The groups associated with the SN location are 1, 1, and 7 from top to bottom. The colors of the lines are matched with the colors of the groups in Fig. 5.2

5.2.3 Delay Time Distribution Models

Delay time distribution (DTD) describes the SN rate of a burst of star formation activity after a given time t. In our research, we used four candidate DTD models. The first model (denoted as MDA) is a simple power-law DTD model with two parameters to be constrained. The SN Ia rate evolution with time in MDA is in the following form:

$$DTD(t) = \begin{cases} 0, & t < \tau \\ A \times B \times t^s, & t \geq \tau \end{cases}, \tag{5.7}$$

where τ is the delay time, s is the slope index, A is the normalization factor for the absolute SN rate which can be calibrated by the SN rates derived from SN surveys, and B is the normalization factor as defined below in Eq. 5.11. In our calculations, τ and s are constrained using maximum likelihood estimate. In the maximum likelihood estimate, we choose a plain prior for $\log(\tau)$ and s: $log_{10}(\tau) \in [7.3, 10]$; $s \in [-6, 0]$.

The second model (denoted as MDB) is a broken power-law DTD model with four parameters. The SN Ia rate is

$$DTD(t) = \begin{cases} 0, & t < \tau \\ A \times B \times t^{s_1}, & t_c \geq t \geq \tau \\ A \times B \times t_c^{s_1 - s_2} t^{s_2}, & t \geq t_c \end{cases}, \tag{5.8}$$

where the symbols A, B, and τ have the same meanings as in Eq. 5.7, t_c is the critical time, and s_1 and s_2 are the slopes for the two power-law components. τ, s_1, s_2, and t_c will be constrained in our calculation. In the maximum likelihood estimates, the priors are $log_{10}(\tau) \in [7.3, 10]$; $s_1 \in [-6, 0]$; $s_2 \in [-6, 0]$; $log_{10}(t_c) \in [9.25, 10.1]$.

The third model (denoted as MDC) is the sum of two power-law relations with the same slope. There are four parameters to be constrained. The SN Ia rate is given by

$$DTD(t) = \begin{cases} 0, & t < \tau \\ A \times B \times t^s, & t_c \geq t \geq \tau \ , \\ A \times B \times r \times t^s, & t \geq t_c \end{cases} \tag{5.9}$$

where the symbols A, B, s, and τ have the same meanings as in Eq. 5.7, t_c is the critical time, and r is the ratio between the two components. In the maximum likelihood method, the priors are $log_{10}(\tau) \in [7.3, 10]$; $s \in [-6, 0]$; $log_{10}(t_c) \in [9.25, 10.1]$; $log_{10}(r) \in [-3, 3]$.

The fourth model (denoted as MDD) is also a combination of two power-law relations, but the slopes are different so the model contains five parameters to be constrained. The SN Ia rate is

$$DTD(t) = \begin{cases} 0, & t < \tau \\ A \times B \times t^{s_1}, & t_c \geq t \geq \tau \ , \\ A \times B \times r \times t_c^{s_1 - s_2} t^{s_2}, & t \geq t_c \end{cases} \tag{5.10}$$

where s_1 and s_2 are the slopes for the two components. In the maximum likelihood estimate, the priors are $log_{10}(\tau) \in [7.3, 9]$; $s_1 \in [-6, 0]$; $s_2 \in [-6, 0]$; $log_{10}(t_c) \in [9.25, 10.1]$; $log_{10}(r) \in [-3, 3]$.

The normalization factors B in all the four DTD models satisfy

$$\int_0^{T_{cosmic}} DTD(t)dt = A, \tag{5.11}$$

where T_{cosmic} is the cosmic age. We set $T_{cosmic} = 17.78$ Gyr to conform with the stellar population grid used in `ppxf`. B is effectively a normalization factor such that the coefficient A is directly determined from observed cosmic supernova rates.

5.2.4 Maximum Likelihood Estimate

The probability of finding an SN at position (x_0, y_0) at the present time can be obtained by integrating the contributions from all stars in the host galaxy and through the cosmic times of their evolution,

$$\begin{aligned} P_{SN}(x_0, y_0) \propto \int_0^{t_{max}} dt' \int\int dxdy \ \Omega(x - x_0, y - y_0) * \\ R(x, y) SPAD(t', x, y) * DTD(t'), \end{aligned} \tag{5.12}$$

where t' is the lookback time, t_{max} is the age of the Universe, $R(x, y)$ is a scaling factor that accounts for the total number of stars formed at position (x, y) and is given by the ratio of the observed and the model ppxf spectrum at the position (x, y), $\Omega(x - x_0, y - y_0)$ is a window function accounting for the probability that the SN at (x_0, y_0) can be related to the stellar groups located at any given (x, y), and $SPAD(t', x, y)$ is the SPAD at position (x, y).

The conditional probability that an SN occurs at position (x_0, y_0) knowing there is an SN from the host galaxy is

$$P_R(x_0, y_0) = \frac{P_{SN}(x_0, y_0)}{\int_0^{t_{max}} dt' \int\int dxdy R(x, y) SPAD(t', x, y) * DTD(t').} \tag{5.13}$$

We adopt a simple Gaussian window function of the form

$$\Omega(x - x_0, y - y_0) = \frac{1}{2\pi\sigma_D^2} exp\left(-\frac{1}{2}\left(\frac{x - x_0}{\sigma_D}\right)^2 - \frac{1}{2}\left(\frac{y - y_0}{\sigma_D}\right)^2\right), \tag{5.14}$$

where σ_D is the Gaussian width whose minimum value is given by the seeing of the observations.

The joint likelihood function for all of the selected SNe (N_{SN}) is

$$L_k = \prod_{l=1}^{N_{SN}} P_{R,l} \ , \tag{5.15}$$

where $P_{R,l}$ is the conditional probability of the l-th SN in the sample calculated using Eq. 5.12.

In our research, we replace the SPAD at each pixel ($SPAD(t', x, y)$) with the SPAD of the relating group, denoted as $SPAD_{\mathcal{I}}(t', x, y)$, where $\mathcal{I}$ is the number of the K subgroups as defined in Sect. 5.2. Therefore, the Eqs. 5.12 and 5.13 are rewritten as

$$P_{SN}(x_0, y_0) \propto \int_0^{t_{max}} dt' \int\int dxdy\ \Omega(x - x_0, y - y_0) * R(x, y) SPAD_{\mathcal{I}}(t', x, y) * DTD(t'), \tag{5.16}$$

$$P_R(x_0, y_0) = \frac{P_{SN}(x_0, y_0)}{\int_0^{t_{max}} dt' \int\int dxdy R(x, y) SPAD_{\mathcal{I}}(t', x, y) * DTD(t').} \tag{5.17}$$

In Sect. 5.3.2, we discuss the difference between the DTD estimate results from Eqs. 5.12 $\sim$ 5.13 and from Eqs. 5.16 $\sim$ 5.17 and the impact of Eq. 5.14's σ_D parameter on the results.

We use the Markov chain Monte Carlo-based code `emcee` [17] for maximum likelihood estimation of the DTD parameters. In Sect. B.1, we use the MUSE data to simulate artificial SN explosion positions from a given DTD model and then test whether our EMD-based algorithm can reproduce the injected DTD model. We found our algorithm can reproduce the DTD model with reasonable accuracy, and we did not observe any bias in the results.

5.3 Results

The MUSE dataset is larger than the MaNGA dataset and is with much higher quality, and we focus our research on the MUSE data, but also provide the results from MaNGA for comparisons. In Sect. 5.3.1, we show the results of the DTD estimates after applying different numbers of groups K to the MUSE data. In Sect. 5.3.2, we discuss the effect of separating galaxies into groups. In Sect. 5.3.3, we discuss the DTD estimates using the MaNGA data. In Sect. 5.3.4, we apply four different DTD models to the MUSE data and compare the model performances using the Bayesian information criterion (BIC).

5.3.1 The Total Number of SPAD Groups

In this section, we investigate the effect of the number of SPAD groups on the constraints of the DTD parameters. In Eq. 5.14, σ_D is set to be the seeing profiles of the observations. In Fig. 5.4, we present the posterior probability distribution of the MDA model with $K = 2, 7, 15$. We notice that the maximum likelihood for all the posterior probability distributions is around $log_{10}(\tau) = 8.2$ and $s = -1.2$, and the 1σ intervals for both τ and s parameters are more tightly constrained with larger values of K.

In Fig. 5.5, we show the 1σ limits, the median absolute deviations, and the median values of $log_{10}(\tau)$ and s estimated with different grouping numbers. The results using K from 2 to 15 are consistent with each other, but with larger fluctuations when K is smaller than 4. It is encouraging that the fluctuations decrease with increasing K values and converge at the higher end. For $K = 15$, the MDA model parameter median and 1σ limit estimates are $\tau = 120^{+142}_{-83} Myr$ and $s = -1.41^{+0.32}_{-0.33}$.

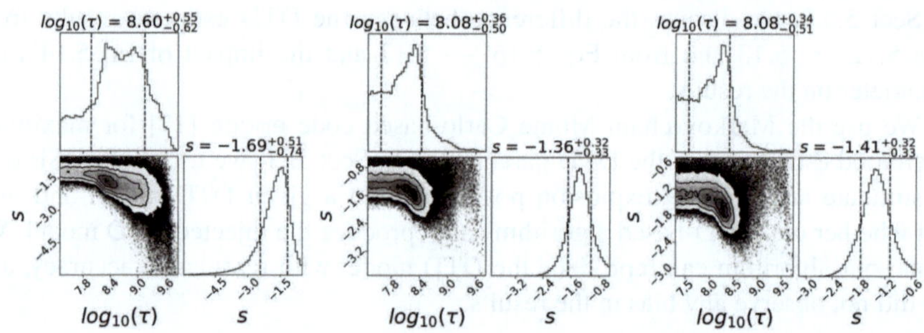

Fig. 5.4 From the left to the right, the posterior probability distributions of the MDA model parameters using $k = 2$, 7, and 15 groups for the MUSE data

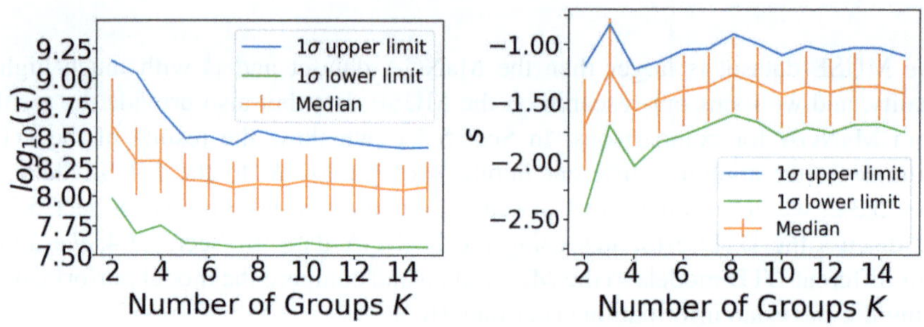

Fig. 5.5 Upper panel: The 1σ upper limit, median, and 1σ lower limit of $log_{10}(\tau)$ with a different number of groups K. Lower panel: The 1σ upper limit, median, and 1σ lower limit of s with a different number of groups K. The error bar is the median absolute deviation

5.3.2 The Effect of Grouping

In Eq. 5.14, we use a two-dimensional Gaussian function to emulate the contribution of SN probability from nearby stellar populations, and the choice of σ_D refers to the seeing profile. In this section, we consider the scenario that the SN progenitor is dynamically segregated from its parent population. Thus, σ_D is written as

$$\sigma_D = \sqrt{\sigma_{seeing}^2 + \sigma_{drift}^2}, \tag{5.18}$$

where σ_{seeing} is the Gaussian width of the seeing of the observation, and σ_{drift} is the projected angular distance between the SN and the progenitor's parent stellar population. We consider σ_{drift} to be 0, 1 kpc, or 5 kpc, and σ_{seeing} comes from the seeing profile. Moreover, we consider an extreme case that $\sigma_D = 0$, which means Ω become a delta function, to further probe the impact of σ_D to the DTD parameter estimate results.

In Fig. 5.6, we show the effect of σ_D on the DTD parameter estimate results, and the SPAD without grouping (Eqs. 5.12 $\sim$ 5.13) or the grouped SPAD with

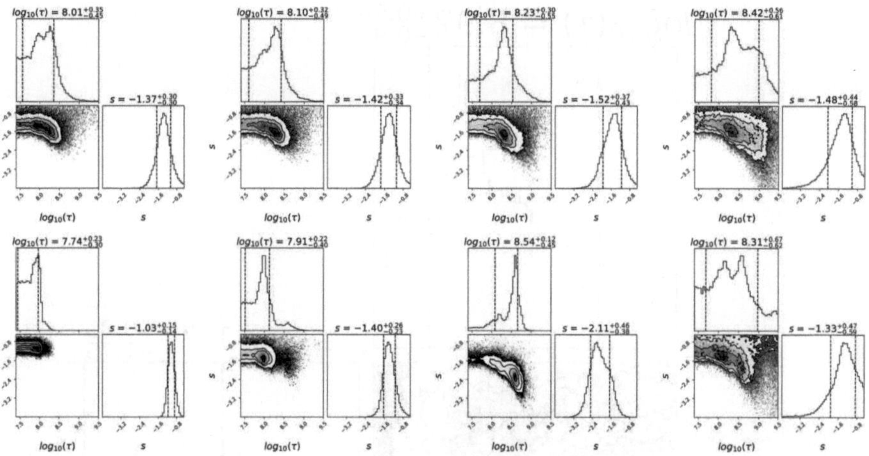

Fig. 5.6 The upper panel shows the posterior probability distributions of the MDA model with the SPAD grouping method (Eqs. 5.16 $\sim$ 5.17 are used), the number of group is $K = 15$, and the lower panel shows that without SPAD grouping (Eqs. 5.12 $\sim$ 5.13 are used). From left to right, the σ_D parameters are: (1) $\sigma_D = 0$. (2) $\sigma_D = \sigma_{seeing}$. (3) $\sigma_D = \sqrt{\sigma_{seeing}^2 + (1\ kpc)^2}$. (4) $\sigma_D = \sqrt{\sigma_{seeing}^2 + (5\ kpc)^2}$

$K = 15$ (Eq. 5.16 $\sim$ 5.17) are used. We notice when the SPADs are grouped, the estimates for $log_{10}(\tau)$ vary between 8.01 and 8.42, and the estimates for s vary between -1.37 and -1.52. In contrast, when the SPADs are not grouped, the estimates for $log_{10}(\tau)$ vary between 7.74 and 8.54, and the estimates for s vary between -1.03 and -2.11. It is not clear whether the change of σ_D can be a good test for the DTD parameter estimates, but we notice the DTD parameters show smaller variations when the grouping method is adopted.

5.3.3 Results from MaNGA

The posterior probability distribution for the DTD model MDA is shown in Fig. 5.7. The number of group is set to 15, and σ_D is the observational seeing. We found the delay time to be $832^{+240}_{-734} Myr$ and the slope $s = -1.77^{+0.80}_{-1.75}$, which are in broad agreement with the values deduced from the MUSE data. We hypothesize the large uncertainties of the DTD parameters derived from MaNGA data is due to the limited data set size: (1) The available SN sample is smaller than that in the MUSE data set; (2) The available spectra in one MaNGA data cube is less than a MUSE data cube. In Appendix B.2, we use the SPAD calculated by `Pipe3D` to estimate the DTD parameters and found the delay time to be 407^{+1252}_{-341}, the slope $s = -1.55^{+0.61}_{-1.01}$, which is consistent with the values using `ppxf` to calculate SPAD. To notice, according to the DTD estimate simulations in [18] and the large error

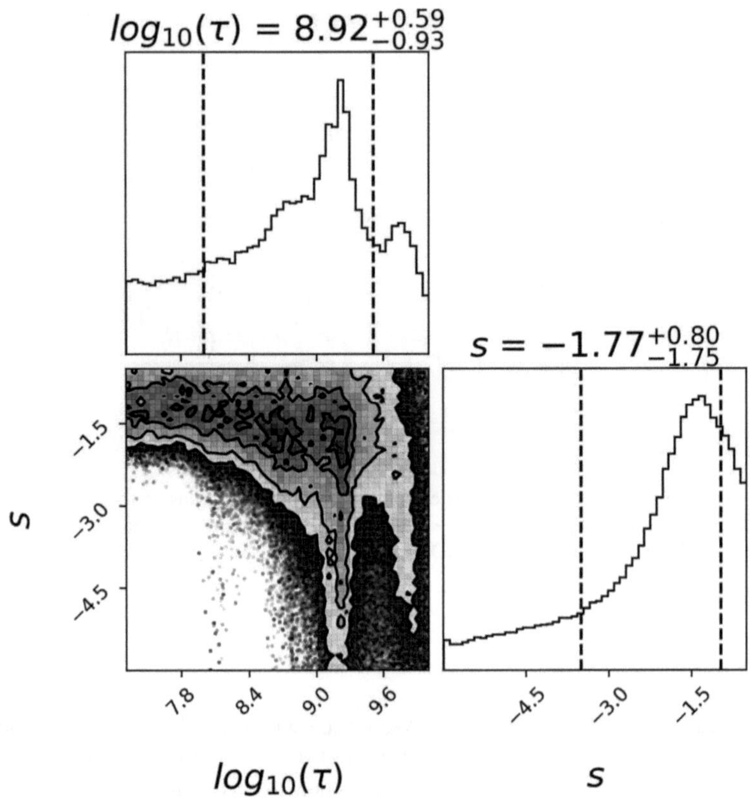

Fig. 5.7 The posterior probability distribution of the MDA model parameters using $K = 15$ groups for the MaNGA data

shown in our τ and s estimates, the result based on 30 MaNGA data cubes may not be reliable and more observations are necessary.

5.3.4 Model Comparison

Based on Sect. 5.3.1, we use Eq. 5.15 as the likelihood function and set $K = 15$ for the parameter constraints of DTD models MDB, MDC, and MDD. The results are shown in Fig. 5.8.

Comparing to the results using the MDA model, the parameters in the MDB, MDC, and MDD models are less constrained, but still show a maximum likelihood at around $log_{10}(\tau) = 8.2$. In the posterior probability distribution of MDB, the two slope parameters s_1 and s_2 are close to the slope s in MDA. Moreover, the critical time t_c for MDB is close to the age of Universe, which questions the necessity of the second component employed in MDB. The parameters in MDC and MDD (s

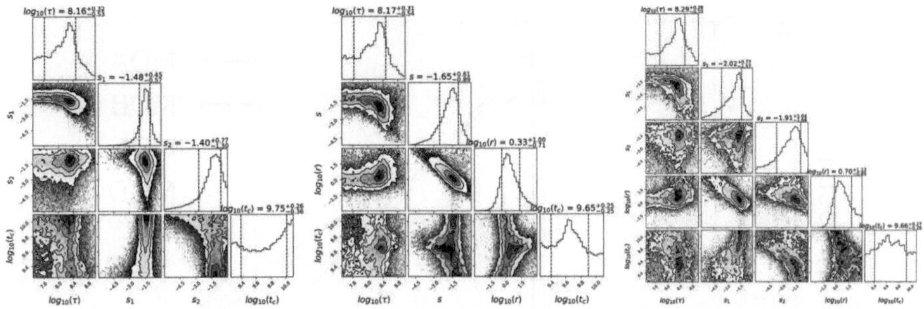

Fig. 5.8 Left: The posterior probability distribution of MDB model parameters using $K = 15$ for the MUSE data. Middle: The posterior probability distribution of MDC model parameters using $K = 15$ for the MUSE data. Right: The posterior probability distribution of MDD model parameters using $K = 15$ for the MUSE data

Table 5.1 The BIC values for the four DTD models

Model name	τ	s or s_1	s_2	r	t_c	k_{dof}	BIC
MDA	$10^{8.275}$	−1.648	None	None	None	2	495.49
MDB	$10^{8.320}$	−2.363	−1.167	None	$10^{9.263}$	4	503.68
MDC	$10^{8.385}$	−2.566	None	$10^{1.069}$	$10^{9.633}$	4	502.77
MDD	$10^{8.409}$	−3.026	−2.061	$10^{1.268}$	$10^{9.577}$	5	507.14

in MDC and s_1 and s_2 in MDD; r in MDC and MDD; t_c in MDC and MDD) are all in agreement to within the statistical errors. In all these models, t_c was found to be around $10^{9.5}$ years. Both MDC and MDD give a value for the ratio r larger than 1, which suggests a population of SNe Ia descend from old stellar populations at around $10^{9.6}$ years albeit with large errors.

To investigate further whether the data can set constraints on progenitor systems with drastically different DTDs, we adopted the Bayesian information criterion (BIC) [19] to assess the goodness of the fits using the different DTD models. The BIC is defined as

$$BIC = k_{dof}\ ln(n) - 2ln(L), \tag{5.19}$$

where k_{dof} is the degree of freedom, n is the size of data sample, and L is the maximized likelihood function. In our research, $n = 100$ as we have 100 SNe Ia coordinates, L is calculated from Eq. 5.15, and k_{dof} are 2, 4, 4, and 5 for MDA, MDB, MDC, and MDD, respectively. In Table 5.1, we show the DTD model parameters which maximize the likelihood and their BIC values. The DTD profiles for these models are shown in Fig. 5.9.

The one-component model MDA shows the smallest BIC value, while all the two-component models show larger BIC values. According to [20], the model with a smaller BIC value is preferred when the BIC value difference of the two models is

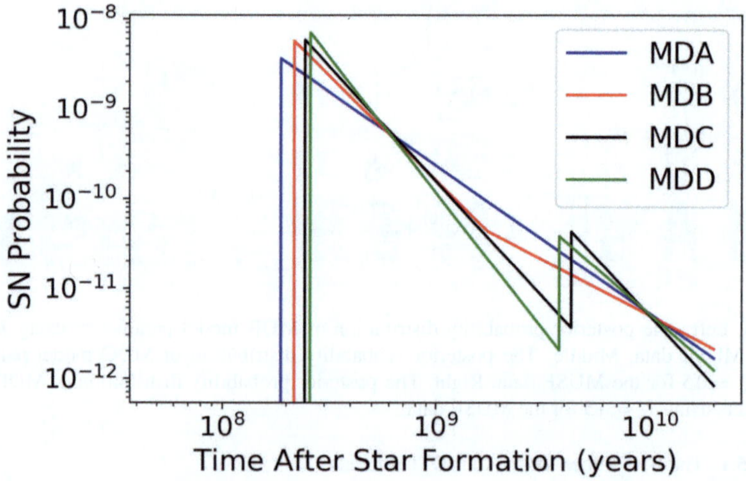

Fig. 5.9 The SN Ia rate for the DTD models MDA (ultramarine), MDB (red), MDC (black), and MDD (green). All the DTDs are normalized so that their time integral is equal to 1. The parameters are shown in Table 5.1

larger than 2. From this test, we conclude that given the current dataset, we cannot establish the existence of multicomponents in the progenitors of SNe Ia.

References

1. Weilbacher PM, Streicher O, Urrutia T, Jarno A, Pécontal-Rousset A, Bacon R, Böhm P (Sept. 2012) Design and capabilities of the MUSE data reduction software and pipeline. In: Software and cyberinfrastructure for astronomy II, vol 8451. Society of photo-optical instrumentation engineers (SPIE) conference series, p 84510B
2. Bundy K, Bershady MA, Law DR, Yan R, Drory N, MacDonald N, Wake DA, Cherinka B, Sánchez-Gallego JR, Weijmans A-M, Thomas D, Tremonti C, Masters K, Coccato L, Diamond-Stanic AM, Aragón-Salamanca A, Avila-Reese V, Badenes C, Falcón-Barroso J, Belfiore F, Bizyaev D, Blanc GA, Bland-Hawthorn J, Blanton MR, Brownstein JR, Byler N, Cappellari M, Conroy C, Dutton AA, Emsellem E, Etherington J, Frinchaboy PM, Fu H, Gunn JE, Harding P, Johnston EJ, Kauffmann G, Kinemuchi K, Klaene MA, Knapen JH, Leauthaud A, Li C, Lin L, Maiolino R, Malanushenko V, Malanushenko E, Mao S, Maraston C, McDermid RM, Merrifield MR, Nichol RC, Oravetz D, Pan K, Parejko JK, Sanchez SF, Schlegel D, Simmons A, Steele O, Steinmetz M, Thanjavur K, Thompson BA, Tinker JL, van den Bosch RCE, Westfall KB, Wilkinson D, Wright S, Xiao T, Zhang K (Jan. 2015) Overview of the SDSS-IV MaNGA survey: mapping nearby galaxies at apache point observatory. Astrophys J 798:7
3. Westfall KB, Cappellari M, Bershady MA, Bundy K, Belfiore F, Ji X, Law DR, Schaefer A, Shetty S, Tremonti CA, Yan R, Andrews BH, Brownstein JR, Cherinka B, Coccato L, Drory N, Maraston C, Parikh T, Sánchez-Gallego JR, Thomas D, Weijmans A-M, Barrera-Ballesteros J, Du C, Goddard D, Li N, Masters K, Ibarra Medel HJ, Sánchez SF, Yang M, Zheng Z, Zhou S (Dec. 2019) The data analysis pipeline for the SDSS-IV MaNGA IFU galaxy survey: overview. Astron J 158:231

4. Cherinka B, Andrews BH, Sánchez-Gallego J, Brownstein J, Argudo-Fernández M, Blanton M, Bundy K, Jones A, Masters K, Law DR, Rowlands K, Weijmans A-M, Westfall K, Yan R (Jul 2019) Marvin: a tool kit for streamlined access and visualization of the SDSS-IV MaNGA data set. Astron J 158:74
5. Bertin E, Arnouts S (June 1996) SExtractor: software for source extraction. Astron Astrophys Suppl 117:393–404
6. Guérou A, Emsellem E, Krajnović D, McDermid RM, Contini T, Weilbacher PM (July 2016) Exploring the mass assembly of the early-type disc galaxy NGC 3115 with MUSE. Astron Astrophys 591:A143
7. Cappellari M (Apr. 2017) Improving the full spectrum fitting method: accurate convolution with Gauss-Hermite functions. Mon Not R Astron Soc 466:798–811
8. Salasnich B, Girardi L, Weiss A, Chiosi C (Sept. 2000) Evolutionary tracks and isochrones for alpha -enhanced stars. Astron Astrophys 361:1023–1035
9. Kroupa P (Jan. 2002) The initial mass function of stars: evidence for uniformity in variable systems. Science 295:82–91
10. McDermid RM, Alatalo K, Blitz L, Bournaud F, Bureau M, Cappellari M, Crocker AF, Davies RL, Davis TA, de Zeeuw PT, Duc P-A, Emsellem E, Khochfar S, Krajnović D, Kuntschner H, Morganti R, Naab T, Oosterloo T, Sarzi M, Scott N, Serra P, Weijmans A-M, Young LM (Apr. 2015) The ATLAS^{3D} Project - XXX. Star formation histories and stellar population scaling relations of early-type galaxies. Mon Not R Astron Soc 448:3484–3513
11. Guérou A, Emsellem E, McDermid RM, Côté P, Ferrarese L, Blakeslee JP, Durrell PR, MacArthur LA, Peng EW, Cuilland re J-C, Gwyn S (May 2015) The next generation virgo cluster survey. XII. Stellar populations and kinematics of compact, low-mass early-type galaxies from gemini GMOS-IFU spectroscopy. Astrophys J 804:70
12. Press WH, Teukolsky SA, Vetterling WT, Flannery BP (1992) Numerical recipes in FORTRAN. The art of scientific computing. Cambridge University Press, Cambridge
13. Rubner Y, Tomasi C, Guibas LJ (Nov. 2000) The earth mover's distance as a metric for image retrieval. Int J Comput Vis 40:99–121
14. Arjovsky M, Chintala S, Bottou L (Jan. 2017) Wasserstein GAN. arXiv e-prints, arXiv:1701.07875
15. Indyk P, Price E (Apr. 2011) K-median clustering, model-based compressive sensing, and sparse recovery for earth mover distance. arXiv e-prints, arXiv:1104.4674
16. Mo D, Duarte MF (2013) Compressive parameter estimation with earth mover's distance via K-median clustering. In: Ville DVD, Goyal VK, Papadakis M (eds) Wavelets and sparsity XV, vol 8858, pp 445–454. International Society for Optics and Photonics, SPIE, Bellingham
17. Foreman-Mackey D, Hogg DW, Lang D, Goodman J (Mar. 2013) emcee: The MCMC Hammer. Publ Astron Soc Pacif 125:306
18. Castrillo A, Ascasibar Y, Galbany L, Sánchez SF, Badenes C, Anderson JP, Kuncarayakti H, Lyman JD, Díaz AI (2020) The delay time distribution of supernovae from integral-field spectroscopy of nearby galaxies. Mon Not R Astron Soc 12:staa3876
19. Wit E, Heuvel Evd, Romeijn J-W (2012) All models are wrong...': an introduction to model uncertainty. Stat Neerlan 66(3):217–236
20. Kass RE, Raftery AE (1995) Bayes factors. J Am Stat Assoc 90(430):773–795

Chapter 6
Concluding Remarks

Abstract We present a novel deep learning framework, AIAI (Artificial Intelligence-based Abundance Inference), for reconstructing the chemical and density structures of Type Ia supernovae (SNe Ia) from synthetic and observational spectra. Using a multiresidual neural network (MRNN) trained on TARDIS-generated spectral models, we predict elemental abundances with high accuracy and apply the method to a sample of 22 observed SNe Ia near B-band maximum light. The model successfully captures correlations between ^{56}Ni abundance and light curve parameters such as stretch and Δm_{15} and reveals the time evolution of ^{56}Ni and ^{56}Co in the ejecta. While predicted B-band magnitudes are systematically brighter than those derived from light curve fitting, this discrepancy is attributed to limitations in the TARDIS radiative transfer assumptions. We also apply our method to construct delay time distributions (DTDs) for SNe Ia using spatially resolved star formation histories from 96 host galaxies observed with VLT/MUSE. A power-law DTD model yields best-fit parameters of a characteristic delay time $\tau = 120^{+142}-83\,\mathrm{Myr}$ and a decay slope $s = -1.41^{+0.32}_{-0.33}$, consistent with previous studies and favoring a double-degenerate progenitor scenario. Our results highlight the potential of combining physically motivated simulations with neural networks to constrain SN explosion physics and progenitor demographics. The methods developed here offer a promising path forward for improving SN Ia cosmology and understanding systematic uncertainties in distance measurements, especially at high redshifts.

6.1 Conclusion on the Elemental Abundance Estimates on SNe Ia

We have developed an AIAI method using the MRNN for the reconstruction of the chemical and density structures of SN Ia models generated using the code TARDIS. With multiresidual modification, we successfully trained and tested the predictive power of the neural network.

The relevance of this study to real observations is explored with the limited amount of observational data. With the elemental abundances predicted from the

X. Chen, *Deep Learning on Type Ia Supernovae*, Springer Theses,
https://doi.org/10.1007/978-3-032-13032-7_6

neural network, we successfully derived model fits to the spectra of SN2011fe, SN2011iv, SN2015F, SN2011by, SN2013dy, ASASSN-14lp, and other 15 SNe Ia near their B-band maximum. The AIAI allows derivations of chemical structures of these SNe.

From the AIAI deduced elements, we found that SNe Ia with higher stretch contains larger Ni mass in Zone 2 and Zone 3. We also observed the decline of the mass of ^{56}Co due to radioactive decay in some well-observed SNe.

We also tried to predict the B-band luminosity from the spectral shapes using AIAI network. The predicted B-band absolute magnitude is systematically higher than the luminosity derived from light curve fits, by $0.47 \sim 1.26$ magnitudes. We surmise the discrepancy is due to approximations of physical processes made in TARDIS.

Despite these successes, we must stress that the present study is only a preliminary exploration of an exciting approach to SN modeling. The study proves that the combination of deep learning techniques with physical models of complicated spectroscopic data may yield unique insights to the physical processes in SNe Ia. However, there are a few caveats that need to be kept in mind, and in a way, these caveats also point to the directions of further improvements:

- In TARDIS, some major assumptions need to be kept in mind. The temperature profiles are calculated based on the assumption that the photospheric spectra follow that of black-body radiation. There are radioactive material very close or above to the photosphere so the energy deposition is significantly more complicated than TARDIS assumptions.
- The current implementation of the code only applies to data around optical maximum.
- We subdivided the ejecta into some artificial grids, and this introduces unphysical boundaries that are inconsistent with the physics of nuclear burning in the ejecta.
- The dependence on the adopted baseline model has not been explored yet, and models with different density and chemical profiles need to be studied and built into the spectral library.
- The spectral models are drawn from a uniform distribution of parameters around the baseline model. This is literally a Bayesian prior. Different Bayesian priors need to be explored.
- The highest ejecta velocity explored in this study is 25,000 km/sec, which may be too low for some high-velocity supernovae such as SN 2004dt [1, 2].
- The size of the spectral library is still tiny.

On a positive note, the current study is sensitive to spectral variations by construction, and the performance of the AIAI network confirms that. We expect the physical approximations made in TARDIS to have only a weak effect on the derivation of chemical elemental abundances.

In the further study, we have developed a deep learning technique to extract physical quantities of supernova spectra. Preliminary application of the methods to a set of observational data proves the method to be powerful. More studies are needed to fully realize the potential of the techniques derived in this study.

We employed the radiative transfer program TARDIS to build a set of NNs to calculate the Ni elemental abundance in the SN Ia ejecta. We measured the Ni abundances and their time evolution in the 13,000–17,000 km/s region of 153 SNe Ia and correlated the Ni abundance to the spectral feature around 3950 Å and the supernova light curve parameter Δm_{15}. The NN prediction on the Ni in the 13,000–17,000 km/s region can reach 0.2 dex accuracy on the simulated SN 2011fe spectra, and the sensitivity can reach $10^{-2.5}$ of Ni mass fraction. The predictive errors are typically smaller than 0.4 dex, and the predicted Ni abundance is correlated to the spectral line feature $FRNi$, which may allow us to measure the amount of Ni in SN ejecta directly. The predicted Ni abundance is also correlated with the light curve parameter Δm_{15}, especially in the early-phase spectra, probably due to more ^{56}Ni present in the early-phase SNe Ia ejecta. The Ni spectral line features $FRNi$ and the ejecta structures predicted by the NNs may be further used in cosmological studies requiring more accurate distance measurements with well-controlled systematic errors.

The simulated spectra based on the NN predictions represent a significant improvement in the quality of the spectral modeling of SNe compared to similar models in the literature. This is obviously because of the optimization procedure involved in the AIAI modeling, which searches and interpolates for the best solution from a large set of theoretical models. However, there are many approximations involved in our models, and most noticeably, they include the approximate NLTE calculation in the radiative transfer simulation program TARDIS, the limitations on constructing a comprehensive training dataset, and the oversimplification of the theoretical SN ejecta geometry and density profiles. Note that the deviation of the TARDIS model from Sedona and CMFGEN is found to be the largest in the UV wavelength range [3]. The TARDIS code generates all its photons from an inner boundary, and the other part of the ejecta only reprocesses photons from the inner boundary. Sedona made the improvement of more realistically generating photons in the Monte Carlo simulation, but Sedona models are much more costly to compute.

AI makes it possible to "calibrate" codes with simplified approximations to make them agree with results from codes with a more complete treatment of physical processes. The physical model-based parameterization of the observational data of SNe Ia as developed in this study can be extended to other radiative transfer codes [4–7]. The physical parameters in the AI models can also be used for refining SNe Ia as distance indicators. The large quantity of these allows for a comprehensive search for the dominant effects on the SN luminosities. This approach is different from what has been done so far based on empirical models of SN Ia luminosity-distance relations [8–14]. For future projects of supernova cosmology, such theoretical model-based parameterization has a different sensitivity to the systematic errors of supernova evolution—AIAI may prove to be helpful in constraining the systematic evolution of the age and metallicity of SN Ia progenitors, especially when very high-redshift ($z \sim 3-6$) SNe Ia are involved [15].

It is also essential to have a fast code to generate surrogate models of 3-D radiative transfer to account for the asymmetric geometry of the ejecta as found from spectropolarimetry observations [16–19] and theoretical models [20].

6.2 Conclusion on the SNe Ia Delay Time Distribution

We selected 96 host galaxies of SNe Ia, most of which are observed by the VLT + MUSE under the AMUSING program[1] to calculate the spatially resolved host galaxy SPADs and to constrain the DTD of the SN progenitors. A statistical method to spatially separate the SFH of a galaxy into multiple groups is developed and applied to constrain the model parameters of the DTD models of SNe Ia. We found that the simple power-law model MDA provides the best fit to the data, with the delay time $\tau = 120^{+142}_{-83}$ Myr, and the SN rate decay slope $s = -1.41^{+0.32}_{-0.33}$. We have not found significant evidence for the multicomponent model of SN progenitors based on our analyses.

The slope parameter has been measured in previous studies. For example, [21] used a revised CSFH and derived SN Ia rates at different redshift bins up to $z \sim 2.25$ and constrained the slope to $s = -1.1^{+0.1}_{-0.1}$. Freundlich and Maoz [22] used the HST to search for SNe Ia in 12 massive galaxy clusters at $z \sim 1.13 - 1.75$ and measured the slope to be $s = -1.09^{+0.15}_{-0.12}$. Maoz et al. [23] derived $s = -1.07 \pm 0.07$ from the SDSS image survey data. However, [24] used a color-luminosity method and an SFH reconstruction method on the same dataset and constrained the slope to be $s = -1.34^{+0.19}_{-0.17}$. In contrast, the delay time τ was measured with large uncertainties in previous studies. Maoz and Badenes [25] analyzed SN Ia remnants in the Magellanic Clouds and found a "prompt" SN Ia population, which explodes within 330 Myr of star formation. Our result on s is consistent with previous researches although with larger uncertainties, while our constraints on τ show a higher confidence level than previous studies. Note also, during the process of this paper, [26] showed DTD constraints based on mean stellar age maps from `Pipe3D` [27] using the same MUSE dataset. The results are $\tau = 50^{+100}_{-35}$ Myr and $s = -1.1 \pm 0.3$ (with 50% confidence interval), which is consistent to our result to within the errors.

When compared to the theoretical models of SNe Ia, our DTD results prefer a DD scenario with CO WD+CO WD as progenitor system (e.g., [28, 29]), which shows a delay time $\tau \sim 10^8 - 10^{8.5}$ years and $s \sim -1$. However, [30] and [31] predict a small (∼5%, in [30]) fraction of "prompt" SN Ia population is formed in less than 100 Myr after star formation for the DD scenario. Due to the limitation of stellar population calculation code which can only calculate the stellar population above 63 Myr, we did not attempt to introduce extra structure in our DTD models to accommodate this prompt population. Moreover, as discussed in several previous studies [29, 31], the observed SN Ia rates are most likely explained by a combination of two or more channels in the context of the BPS. Although our result indicates a high confidence of $\tau \sim 120$ Myr, we cannot firmly establish or eliminate the dominant channels of SN Ia progenitors.

[1] https://amusing-muse.github.io/

We failed to derive any strong constraints using the data from MaNGA. However, with the future data releases of SDSS-MaNGA survey program, a detailed analysis on the DTD may be worthwhile using the increased sample size.

In summary, we have developed a new method to estimate the DTD of SNe Ia which allows us to set constraints on the DTD model parameters. With data that we can expect from future observation programs with LSST [32], DESI [33], and HETDEX [34], we may expect this method to be applicable to other subtypes of SNe Ia and core-collapse SNe to set strong constraints on the SN progenitors.

References

1. Wang L, Baade D, Höflich P, Wheeler JC, Kawabata K, Khokhlov A, Nomoto K, Patat F (Dec 2006) Premaximum spectropolarimetry of the type Ia SN 2004dt. Astrophys J 653:490–502
2. Wang X, Filippenko AV, Ganeshalingam M, Li W, Silverman JM, Wang L, Chornock R, Foley RJ, Gates EL, Macomber B, Serduke FJD, Steele TN, Wong DS (July 2009) Improved distances to type Ia supernovae with two spectroscopic subclasses. Astrophys J Lett 699:L139–L143
3. Blondin S, Blinnikov S, Callan FP, Collins CE, Dessart L, Even W, Flörs A, Fullard AG, Hillier DJ, Jerkstrand A, Kasen D, Katz B, Kerzendorf W, Kozyreva A, O'Brien J, Pássaro EA, Roth N, Shen KJ, Shingles L, Sim SA, Singhal J, Smith IG, Sorokina E, Utrobin VP, Vogl C, Williamson M, Wollaeger R, Woosley SE, Wygoda N (Dec. 2022) StaNdaRT: a repository of standardised test models and outputs for supernova radiative transfer. Astron Astrophys 668:A163
4. Hoeflich P, Khokhlov A, Wheeler JC, Phillips MM, Suntzeff NB, Hamuy M (Dec. 1996) Maximum brightness and postmaximum decline of light curves of type IA supernovae: a comparison of theory and observations. Astrophys J Lett 472:L81
5. Hillier DJ, Miller DL (Mar. 1998) The treatment of non-LTE line blanketing in spherically expanding outflows. Astrophys J 496:407–427
6. Baron E, Hauschildt PH (Mar. 1998) Parallel implementation of the PHOENIX generalized stellar atmosphere program. II. Wavelength parallelization. Astrophys J 495:370–376
7. Kasen D, Thomas RC, Nugent P (Nov. 2006) Time-dependent Monte Carlo radiative transfer calculations for three-dimensional supernova spectra, light curves, and polarization. Astrophys J 651:366–380
8. Phillips MM (Aug. 1993) The absolute magnitudes of type IA supernovae. Astrophys J Lett 413:L105
9. Tripp R (Mar. 1998) A two-parameter luminosity correction for Type IA supernovae. Astron Astrophys 331:815–820
10. Jha S, Riess AG, Kirshner RP (Apr. 2007) Improved distances to type Ia supernovae with multicolor light-curve shapes: MLCS2k2. Astrophys J 659:122–148
11. Guy J, Astier P, Nobili S, Regnault N, Pain R (Dec. 2005) SALT: a spectral adaptive light curve template for type Ia supernovae. Astron Astrophys 443:781–791
12. Wang L, Goldhaber G, Aldering G, Perlmutter S (Jun 2003) Multicolor light curves of type Ia supernovae on the color-magnitude diagram: a novel step toward more precise distance and extinction estimates. Astrophys J 590:944
13. Wang L, Strovink M, Conley A, Goldhaber G, Kowalski M, Perlmutter S, Siegrist J (Apr 2006) Nonlinear decline-rate dependence and intrinsic variation of type Ia supernova luminosities. Astrophys J 641:50
14. Wang X, Filippenko AV, Ganeshalingam M, Li W, Silverman JM, Wang L, Chornock R, Foley RJ, Gates EL, Macomber B, Serduke FJD, Steele TN, Wong DS (Jun 2009) Improved distances to type Ia supernovae with two spectroscopic subclasses. Astrophys J 699:L139

15. Lu J, Wang L, Chen X, Rubin D, Perlmutter S, Baade D, Mould J, Vinko J, Regős E, Koekemoer AM (Dec. 2022) Constraints on cosmological parameters with a sample of type Ia supernovae from JWST. Astrophys J 941:71
16. Wang L, Wheeler JC, Li Z, Clocchiatti A (Aug. 1996) Broadband polarimetry of supernovae: SN 1994D, SN 1994Y, SN 1994ae, SN 1995D, and SN 1995H. Astrophys J 467:435
17. Wang L, Wheeler JC (Sept. 2008) Spectropolarimetry of supernovae. Ann Rev Astron Astrophys 46:433–474
18. Cikota A, Patat F, Wang L, Wheeler JC, Bulla M, Baade D, Höflich P, Cikota S, Clocchiatti A, Maund JR, Stevance HF, Yang Y (Nov. 2019) Linear spectropolarimetry of 35 Type Ia supernovae with VLT/FORS: an analysis of the Si II line polarization. Mon Not R Astron Soc 490:578–599
19. Yang Y, Yan H, Wang L, Wheeler JC, Baade D, Isaacson H, Cikota A, Maund JR, Hoeflich P, Patat F, Giacalone S, Rice M, Tyler DB, Mishra D, Ashall C, Brink TG, Filippenko AV, Galbany L, Patra KC, Shahbandeh M, Vasylyev SS, Vinkó J (Nov. 2022) Spectropolarimetry of the thermonuclear supernova SN 2021rhu: high calcium polarization 79 days after peak luminosity. Astrophys J 939:18
20. Neopane S, Bhargava K, Fisher R, Ferrari M, Yoshida S, Toonen S, Bravo E (Jan. 2022) Near-Chandrasekhar-mass type Ia supernovae from the double-degenerate channel. Astrophys J 925:92
21. Maoz D, Graur O (Oct 2017) Star formation, supernovae, iron, and α: consistent cosmic and galactic histories. Astrophys J 848:25
22. Freundlich J, Maoz D (Apr. 2021) The delay time distribution of Type-Ia supernovae in galaxy clusters: the impact of extended star-formation histories. Mon Not R Astron Soc 502:5882–5895
23. Maoz D, Mannucci F, Brandt TD (2012) The delay-time distribution of type Ia supernovae from Sloan II. Mon Not R Astron Soc 426:3282–3294
24. Heringer E, Pritchet C, van Kerkwijk MH (Sep 2019) The delay times of type Ia supernova. Astrophys J 882:52
25. Maoz D, Badenes C (2010) The supernova rate and delay time distribution in the Magellanic Clouds. Mon Not R Astron Soc 407:1314–1327
26. Castrillo A, Ascasibar Y, Galbany L, Sánchez SF, Badenes C, Anderson JP, Kuncarayakti H, Lyman JD, Díaz AI (2020) The delay time distribution of supernovae from integral-field spectroscopy of nearby galaxies. Mon Not R Astron Soc staa3876
27. SF Sánchez, Pérez E, Sánchez-Blázquez P, García-Benito R, Ibarra-Mede HJ, González JJ, Rosales-Ortega FF, Sánchez-Menguiano L, Ascasibar Y, Bitsakis T, Law D, Cano-Díaz M, López-Cobá C, Marino RA, Gil de Paz A, López-Sánchez AR, Barrera-Ballesteros J, Galbany L, Mast D, Abril-Melgarejo V, Roman-Lopes A (Apr. 2016) Pipe3D, a pipeline to analyze Integral Field Spectroscopy Data: II. Analysis sequence and CALIFA dataproducts. Revista Mexicana de Astronomía y Astrofísica 52:171–220
28. Chen X, Jeffery CS, Zhang X, Han Z (Aug. 2012) Some constraints on the lower mass limit for double-degenerate progenitors of type Ia supernovae. Astrophys J Lett 755:L9
29. Liu D, Wang B, Han Z (2018) The double-degenerate model for the progenitors of Type Ia supernovae. Mon Not R Astron Soc 473:5352–5361
30. Ruiter AJ, Belczynski K, Fryer C (July 2009) Rates and delay times of type Ia supernovae. Astrophys J 699:2026–2036
31. Mennekens N, Vanbeveren D, De Greve JP, De Donder E (June 2010) The delay-time distribution of type Ia supernovae: a comparison between theory and observation. Astron Astrophys 515:A89
32. Ivezić Ž, Kahn SM, Tyson JA, Abel B, Acosta E, Allsman R, Alonso D, AlSayyad Y, Anderson SF, Andrew J et al (Mar. 2019) LSST: from science drivers to reference design and anticipated data products. Astrophys J 873:111
33. DESI Collaboration, Aghamousa A, Aguilar J, Ahlen S, Alam S, Allen LE, Allende Prieto C, Annis J, Bailey S, Balland C, Ballester O, Baltay C, Beaufore L, Bebek C, Beers TC, Bell EF, Bernal JL, Besuner R, Beutler F, Blake C, Bleuler H, Blomqvist M, Blum R, Bolton AS,

Briceno C, Brooks D, Brownstein JR, Buckley-Geer E, Burden A, Burtin E, Busca NG, Cahn RN, Cai Y-C, Cardiel-Sas L, Carlberg RG, Carton P-H, Casas R, Castander FJ, Cervantes-Cota JL, Claybaugh TM, Close M, Coker CT, Cole S, Comparat J, Cooper AP, Cousinou MC, Crocce M, Cuby J-G, Cunningham DP, Davis TM, Dawson KS, de la Macorra A, De Vicente J, Delubac T, Derwent M, Dey A, Dhungana G, Ding Z, Doel P, Duan YT, Ealet A, Edelstein J, Eftekharzadeh S, Eisenstein DJ, Elliott A, Escoffier S, Evatt M, Fagrelius P, Fan X, Fanning K, Farahi A, Farihi J, Favole G, Feng Y, Fernandez E, Findlay JR, Finkbeiner DP, Fitzpatrick MJ, Flaugher B, Flender S, Font-Ribera A, Forero-Romero JE, Fosalba P, Frenk CS, Fumagalli M, Gaensicke BT, Gallo G, Garcia-Bellido J, Gaztanaga E, Pietro Gentile Fusillo N, Gerard T, Gershkovich I, Giannantonio T, Gillet D, Gonzalez-de-Rivera G, Gonzalez-Perez V, Gott S, Graur O, Gutierrez G, Guy J, Habib S, Heetderks H, Heetderks I, Heitmann K, Hellwing WA, Herrera DA, Ho S, Holland S, Honscheid K, Huff E, Hutchinson TA, Huterer D, Hwang HS, Illa Laguna JM, Ishikawa Y, Jacobs D, Jeffrey N, Jelinsky P, Jennings E, Jiang L, Jimenez J, Johnson J, Joyce R, Jullo E, Juneau S, Kama S, Karcher A, Karkar S, Kehoe R, Kennamer N, Kent S, Kilbinger M, Kim AG, Kirkby D, Kisner T, Kitanidis E, Kneib J-P, Koposov S, Kovacs E, Koyama K, Kremin A, Kron R, Kronig L, Kueter-Young A, Lacey CG, Lafever R, Lahav O, Lambert A, Lampton M, Landriau M, Lang D, Lauer TR, Le Goff J-M, Le Guillou L, Le Van Suu A, Lee JH, Lee S-J, Leitner D, Lesser M, Levi ME, L'Huillier B, Li B, Liang M, Lin H, Linder E, Loebman SR, Lukić Z, Ma J, MacCrann N, Magneville C, Makarem L, Manera M, Manser CJ, Marshall R, Martini P, Massey R, Matheson T, McCauley J, McDonald P, McGreer ID, Meisner A, Metcalfe N, Miller TN, Miquel R, Moustakas J, Myers A, Naik M, Newman JA, Nichol RC, Nicola A, Nicolati da Costa L, Nie J, Niz G, Norberg P, Nord B, Norman D, Nugent P, O'Brien T, Oh M, Olsen AG K, Padilla C, Padmanabhan H, Padmanabhan N, Palanque-Delabrouille N, Palmese A, Pappalardo D, Pâris I, Park C, Patej A, Peacock JA, Peiris HV, Peng X, Percival WJ, Perruchot S, Pieri MM, Pogge R, Pollack JE, Poppett C, Prada F, Prakash A, Probst RG, Rabinowitz D, Raichoor A, Ree CH, Refregier A, Regal X, Reid B, Reil K, Rezaie M, Rockosi CM, Roe N, Ronayette S, Roodman A, Ross AJ, Ross NP, Rossi G, Rozo E, Ruhlmann-Kleider V, Rykoff ES, Sabiu C, Samushia L, Sanchez E, Sanchez J, Schlegel DJ, Schneider M, Schubnell M, Secroun A, Seljak U, Seo H-J, Serrano S, Shafieloo A, Shan H, Sharples R, Sholl MJ, Shourt WV, Silber JH, Silva DR, Sirk MM, Slosar A, Smith A, Smoot GF, Som D, Song Y-S, Sprayberry D, Staten R, Stefanik A, Tarle G, Sien Tie S, Tinker JL, Tojeiro R, Valdes F, Valenzuela O, Valluri M, Vargas-Magana M, Verde L, Walker AR, Wang J, Wang Y, Weaver BA, Weaverdyck C, Wechsler RH, Weinberg DH, White M, Yang Q, Yeche C, Zhang T, Zhao G-B, Zheng Y, Zhou X, Zhou Z, Zhu Y, Zou H, Zu Y (Oct. 2016) The DESI experiment part I: science, targeting, and survey design. arXiv e-prints, arXiv:1611.00036

34. Hill GJ, Gebhardt K, Komatsu E, Drory N, MacQueen PJ, Adams J, Blanc GA, Koehler R, Rafal M, Roth MM, Kelz A, Gronwall C, Ciardullo R, Schneider DP (Oct. 2008) The Hobby-Eberly Telescope Dark Energy Experiment (HETDEX): description and early pilot survey results. In: Kodama T, Yamada T, Aoki K (eds) Panoramic views of galaxy formation and evolution, vol 399. Astronomical Society of the Pacific Conference Series, p 115

Appendix A
List of All Supernovae and Host Galaxies

Table A.1 lists the 100 SNe Ia in 96 host galaxies observed by VLT observation programs. Table A.2 lists the 30 SNe Ia in 30 host galaxies observed by MaNGA facility in SDSS-DR15.

X. Chen, *Deep Learning on Type Ia Supernovae*, Springer Theses,
https://doi.org/10.1007/978-3-032-13032-7

Table A.1 The SN Ia host galaxies observed by MUSE. SN Name: The names of SNe Ia. SN RA: Right ascension of SN coordinates. SN DEC: Declination of SN coordinates. IFU RA: Right ascension of IFU data-cube center. IFU DEC: Declination of IFU data-cube center. SN time: SN discovery date. IFU time: MUSE observation date. Redshift: SN-host galaxy redshift. ARCFILE: The filename of IFU data-cube product stored in ESO archive, to notice the timestamp is not the IFU observation time

SN name	SN RA	SN DEC	IFU RA	IFU DEC	SN time	IFU time	Redshift	ARCFILE
SN2019fkq	359.1010	−29.0230	359.1016	−29.0238	2019-05-14	2019-09-07	0.0450	ADP.2019-10-07T17:13:49.969
SN2018ezx	62.0326	−8.8313	62.0337	−8.8332	2018-08-12	2016-11-08	0.0329	ADP.2017-01-18T15:19:35.833
SN2018djd	33.6398	−0.7664	33.6410	−0.7658	2018-07-12	2017-10-28	0.0264	ADP.2017-11-20T17:51:27.822
SN2018zz	210.9113	−33.9780	210.9126	−33.9786	2018-03-03	2015-08-08	0.0138	ADP.2016-07-12T07:52:45.162
SN2017hgz	327.0808	−34.9516	327.0838	−34.9529	2017-10-10	2015-10-14	0.0162	ADP.2016-08-08T10:10:02.297
SN2017dps	204.1639	−33.9658	204.1668	−33.9670	2017-05-01	2016-04-11	0.0125	ADP.2017-12-18T14:37:20.881
SN2017cze	167.4450	−13.3807	167.4451	−13.3807	2017-04-11	2016-01-05	0.0149	ADP.2016-07-26T12:48:39.617
SN2016gfk	18.5323	−32.6519	18.5270	−32.6572	2016-09-11	2016-05-20	0.0120	ADP.2016-10-05T16:09:44.597
SN2016aew	212.8595	1.2867	212.8604	1.2860	2016-02-12	2014-06-24	0.0250	ADP.2016-08-02T10:23:14.733
SN2014dm	62.0326	−8.8313	62.0297	−8.8270	2014-09-27	2016-11-08	0.0337	ADP.2017-01-18T15:19:35.833
SN2014at	326.5628	−46.5188	326.5618	−46.5225	2014-04-20	2015-05-30	0.0325	ADP.2016-06-17T18:47:32.957
SN2014ao	128.6391	−2.5461	128.6388	−2.5434	2014-04-17	2019-03-20	0.0139	ADP.2019-04-10T17:46:49.603
SN2013hk	45.5462	15.9276	45.5452	15.9274	2013-12-04	2015-12-27	0.0170	ADP.2016-09-23T00:56:03.893
SN2013fz	65.9446	−51.5998	65.9435	−51.5962	2013-11-02	2015-08-03	0.0206	ADP.2016-07-12T07:27:20.835
SN2013fy	324.3678	−47.0357	324.3630	−47.0319	2013-10-25	2015-06-21	0.0314	ADP.2016-06-25T11:26:36.800
SN2013ef	28.8417	6.6120	28.8363	6.6098	2013-07-04	2015-12-28	0.0172	ADP.2017-06-06T17:14:14.070
SN2013dl	19.6732	−7.4494	19.6740	−7.4444	2013-06-17	2016-01-06	0.0024	ADP.2016-07-26T15:11:37.761
SN2013da	206.4018	−7.3259	206.4009	−7.3257	2013-06-05	2017-04-01	0.0246	ADP.2017-04-11T12:41:12.920
SN2013az	84.9729	−40.5124	84.9672	−40.5078	2013-03-24	2015-09-05	0.0373	ADP.2016-07-25T12:08:30.833
SN2013M	209.9903	−37.8637	209.9862	−37.8637	2013-01-20	2017-04-18	0.0350	ADP.2017-12-12T14:16:55.449
SN2012he	75.2111	−38.6544	75.2086	−38.6532	2012-11-22	2017-08-03	0.0576	ADP.2017-09-22T09:52:40.531
SN2012hd	18.5323	−32.6519	18.5311	−32.6521	2012-11-20	2016-05-20	0.0120	ADP.2016-10-05T16:09:44.597
SN2012gm	349.4017	14.0011	349.4043	14.0025	2012-11-19	2015-06-26	0.0148	ADP.2016-06-25T12:05:07.015

SN2012fw	315.4961	−48.2737	315.4958	−48.2739	2012-08-19	2016-04-13	0.0186	ADP.2016-09-21T13:42:23.565
SN2012et	355.6618	27.0922	355.6618	27.0921	2012-09-12	2016-06-09	0.0249	ADP.2016-09-29T20:36:01.636
SN2011jh	191.8143	−10.0621	191.8101	−10.0631	2011-12-22	2019-02-20	0.0078	ADP.2019-03-07T06:28:16.831
SN2011iy	197.2428	−15.5177	197.2433	−15.5178	2011-12-09	2016-05-12	0.0041	ADP.2016-09-29T05:21:54.104
SN2011iv	54.7145	−35.5881	54.7140	−35.5922	2011-12-02	2017-11-22	0.0065	ADP.2017-12-13T01:47:07.213
SN2010jo	14.3960	−1.3909	14.3982	−1.3926	2010-11-06	2017-07-20	0.0452	ADP.2017-09-11T14:28:03.988
SN2010ev	156.3703	−39.8282	156.3708	−39.8309	2010-06-27	2016-04-13	0.0092	ADP.2016-09-21T13:42:23.507
SN2010dl	323.7516	−0.5111	323.7540	−0.5133	2010-05-24	2017-08-04	0.0302	ADP.2017-09-22T10:41:54.300
SN2010aa	27.1749	−48.6480	27.1800	−48.6502	2010-02-09	2018-05-26	0.0207	ADP.2018-06-02T02:30:18.655
SN2009jr	306.6078	2.9102	306.6085	2.9092	2009-10-08	2017-07-30	0.0166	ADP.2017-09-20T13:08:51.796
SN2009iw	88.8664	−76.9201	88.8568	−76.9211	2009-09-15	2015-09-24	0.0160	ADP.2016-07-28T11:30:23.655
SN2009fk	341.1015	−0.1615	341.0996	−0.1617	2009-05-29	2017-08-04	0.0162	ADP.2017-09-22T10:41:54.308
SN2009ds	177.2708	−9.7303	177.2671	−9.7291	2009-04-28	2016-06-30	0.0192	ADP.2017-10-16T10:25:08.202
SN2009aa	170.9220	−22.2711	170.9262	−22.2707	2009-02-03	2015-04-07	0.0281	ADP.2016-06-09T16:16:30.539
SN2009Y	220.5998	−17.2527	220.5994	−17.2468	2009-02-01	2016-04-03	0.0095	ADP.2016-09-07T10:11:23.531
SN2009I	41.2915	−4.7106	41.2933	−4.7137	2009-01-13	2015-07-21	0.0262	ADP.2016-07-11T15:14:15.422
SN2008ia	132.6464	−61.2779	132.6465	−61.2779	2008-12-07	2016-04-19	0.0217	ADP.2016-09-22T21:00:32.919
SN2008fu	45.6195	−24.4555	45.6188	−24.4560	2008-09-25	2018-07-24	0.0524	ADP.2018-09-11T21:30:28.561
SN2008fl	294.1897	−37.5535	294.1868	−37.5513	2008-09-07	2018-05-27	0.0199	ADP.2018-06-02T03:35:25.145
SN2008ec	345.8151	8.8741	345.8190	8.8722	2008-07-14	2014-08-19	0.0159	ADP.2016-07-14T14:17:17.765
SN2008dh	8.7973	23.2545	8.7972	23.2542	2008-06-08	2016-07-18	0.0368	ADP.2016-10-14T08:21:03.084
SN2008cf	211.8831	−26.5516	211.8857	−26.5518	2008-05-04	2015-05-23	0.0471	ADP.2016-06-17T17:51:10.735
SN2008cc	315.8740	−67.1810	315.8734	−67.1836	2008-04-24	2018-05-27	0.0106	ADP.2018-06-02T03:35:25.161
SN2008bq	100.2658	−38.0356	100.2605	−38.0386	2008-04-02	2018-09-26	0.0346	ADP.2018-10-25T08:36:03.111
SN2008bd	154.5978	−13.1038	154.5972	−13.1031	2008-03-13	2019-03-26	0.0306	ADP.2019-04-17T22:54:16.942
SN2008ar	186.1585	10.8393	186.1580	10.8382	2008-02-27	2015-05-30	0.0262	ADP.2016-06-17T18:47:32.895

(continued)

Table A.1 (continued)

SN name	SN RA	SN DEC	IFU RA	IFU DEC	SN time	IFU time	Redshift	ARCFILE
SN2007st	27.1749	−48.6480	27.1770	−48.6494	2007-12-22	2018-05-26	0.0214	ADP.2018-06-02T02:30:18.655
SN2007so	41.9318	13.2556	41.9297	13.2541	2007-12-13	2015-07-24	0.0298	ADP.2016-07-11T15:28:11.087
SN2007hx	31.6127	−0.8992	31.6128	−0.8995	2007-09-03	2015-07-23	0.0798	ADP.2016-07-11T15:19:32.583
SN2007cq	333.6697	5.0787	333.6685	5.0803	2007-06-21	2017-08-04	0.0263	ADP.2017-09-22T10:41:54.323
SN2007cg	201.3917	−24.6520	201.3899	−24.6522	2007-05-11	2015-05-28	0.0331	ADP.2016-06-17T18:13:44.235
SN2007bc	169.8142	20.8138	169.8107	20.8090	2007-04-04	2015-05-29	0.0208	ADP.2016-06-17T18:25:05.260
SN2007al	149.8290	−19.4729	149.8270	−19.4738	2007-03-10	2015-05-31	0.0122	ADP.2017-03-28T14:09:36.373
SN2007ai	243.2228	−21.6266	243.2239	−21.6302	2007-03-06	2018-05-26	0.0330	ADP.2018-06-02T02:30:18.663
SN2007S	150.1291	4.4072	150.1302	4.4073	2007-01-29	2015-06-26	0.0139	ADP.2016-06-25T12:12:49.422
SN2006os	43.7525	16.0126	43.7542	16.0097	2006-11-21	2015-07-21	0.0328	ADP.2016-07-11T15:14:15.414
SN2006ob	27.9522	0.2636	27.9505	0.2634	2006-11-13	2015-07-13	0.0592	ADP.2016-07-11T14:04:30.018
SN2006lu	138.8208	−25.5999	138.8235	−25.6001	2006-10-30	2015-04-13	0.0540	ADP.2016-06-14T09:15:58.860
SN2006hx	18.4876	0.3719	18.4888	0.3717	2006-09-28	2015-06-21	0.0454	ADP.2016-06-25T11:26:36.852
SN2006hb	75.5042	−21.1342	75.5053	−21.1320	2006-09-27	2018-08-29	0.0153	ADP.2018-10-17T14:54:22.224
SN2006et	10.6911	−23.5616	10.6909	−23.5584	2006-09-03	2015-06-27	0.0223	ADP.2016-06-25T12:12:49.414
SN2006ej	9.7512	−9.0149	9.7490	−9.0157	2006-08-23	2015-06-18	0.0203	ADP.2016-06-24T11:45:22.441
SN2006cm	320.0731	−1.6842	320.0728	−1.6841	2006-05-24	2016-05-19	0.0163	ADP.2016-12-02T09:39:19.087
SN2006br	202.5085	13.4164	202.5075	13.4158	2006-04-25	2015-06-04	0.0247	ADP.2016-06-24T10:20:34.507
SN2006D	193.1445	−9.7772	193.1414	−9.7752	2006-01-11	2015-05-23	0.0086	ADP.2016-06-17T17:51:10.811
SN2005na	105.4042	14.1366	105.4026	14.1332	2005-12-31	2015-04-11	0.0263	ADP.2016-06-21T00:31:05.284
SN2005lu	39.0168	−17.2638	39.0155	−17.2639	2005-12-11	2015-06-18	0.0327	ADP.2016-06-24T11:45:22.429
SN2005ku	344.9251	−0.0134	344.9275	−0.0137	2005-11-10	2015-05-30	0.0454	ADP.2016-06-17T18:47:32.872
SN2005iq	359.6342	−18.7111	359.6354	−18.7092	2005-11-05	2018-07-25	0.0346	ADP.2018-09-13T01:03:07.237
SN2005hc	29.2022	−0.2122	29.1998	−0.2137	2005-10-12	2015-08-02	0.0459	ADP.2016-07-12T07:19:14.082
SN2005bs	302.5615	−56.6390	302.5588	−56.6454	2005-04-19	2016-05-13	0.0552	ADP.2016-09-29T08:33:33.612
SN2005bg	184.3220	16.3717	184.3216	16.3716	2005-03-28	2015-05-29	0.0230	ADP.2016-06-17T18:47:32.927

SN2005be	224.8876	16.6699	224.8863	16.6699	2005-04-05	2015-05-30	0.0336	ADP.2016-06-17T18:47:32.864
SN2005al	207.5037	−30.5772	207.5014	−30.5762	2005-02-24	2018-05-10	0.0124	ADP.2018-05-18T04:03:44.301
SN2005ag	224.1793	9.3286	224.1819	9.3285	2005-02-10	2015-05-29	0.0797	ADP.2016-06-17T18:25:05.248
SN2004gc	80.4543	6.6794	80.4581	6.6760	2004-11-18	2019-02-24	0.0305	ADP.2019-03-08T05:11:20.934
SN2004ey	327.2793	0.4473	327.2825	0.4442	2004-10-14	2015-06-05	0.0158	ADP.2016-06-24T10:28:35.821
SN2004ef	340.5458	19.9971	340.5418	19.9946	2004-09-04	2015-06-05	0.0310	ADP.2016-06-24T10:28:35.837
SN2004do	283.8988	−53.7239	283.8905	−53.7230	2004-08-04	2015-10-07	0.0086	ADP.2016-08-02T05:17:36.565
SN2004cs	267.5579	14.2868	267.5599	14.2832	2004-06-23	2016-03-09	0.0141	ADP.2016-08-17T10:26:59.052
SN2003ic	10.4605	−9.3035	10.4593	−9.3053	2003-09-16	2018-08-10	0.0554	ADP.2018-09-20T04:52:51.377
SN2003gh	116.3256	−71.4095	116.3247	−71.4104	2003-06-29	2017-12-02	0.0179	ADP.2017-12-20T14:19:23.117
SN2002jg	334.8675	29.3897	334.8700	29.3846	2002-11-23	2016-05-25	0.0162	ADP.2016-10-07T07:11:23.283
SN2002fk	50.5283	−15.3994	50.5238	−15.4009	2002-09-17	2015-09-10	0.0071	ADP.2016-07-25T12:56:04.451
SN2001da	358.3843	8.1183	358.3866	8.1174	2001-07-09	2016-05-22	0.0172	ADP.2016-10-05T16:52:03.490
SN2001E	177.2708	−9.7303	177.2708	−9.7364	2001-01-05	2016-06-30	0.0192	ADP.2017-10-16T10:25:08.202
SN2001A	184.8467	5.8251	184.8459	5.8279	2001-01-01	2016-04-17	0.0073	ADP.2016-09-22T13:48:58.454
SN2000fs	47.1098	4.1109	47.1093	4.1111	2000-09-06	2018-11-03	0.0300	ADP.2018-11-12T14:39:55.596
SN2000do	287.8565	−50.6404	287.8591	−50.6401	2000-09-30	2017-06-18	0.0109	ADP.2018-08-02T18:48:39.520
SN2000A	351.9877	8.7785	351.9954	8.7839	2000-01-01	2017-09-15	0.0296	ADP.2017-10-06T15:09:34.514
SN1999ee	334.0384	−36.8439	334.0417	−36.8444	1999-10-07	2014-10-27	0.0114	ADP.2016-06-23T09:51:35.962
SN1998V	275.6593	15.6966	275.6558	15.7023	1998-03-10	2019-05-06	0.1753	ADP.2019-07-20T08:00:23.865
SN1997dt	345.0154	15.9802	345.0122	15.9808	1997-11-22	2018-06-23	0.0073	ADP.2018-08-09T21:08:02.554
SN1994D	188.5094	7.7007	188.5102	7.7013	1994-03-07	2016-05-26	0.0027	ADP.2016-10-06T16:52:10.915
SN1991S	157.3645	22.0083	157.3658	22.0129	1991-04-10	2017-01-20	0.0544	ADP.2017-03-20T10:48:00.633
SN1989B	170.0563	13.0061	170.0579	13.0053	1989-01-30	2018-05-14	0.0023	ADP.2018-05-29T18:17:29.607
SN1968I	197.2036	−6.7776	197.2054	−6.7778	1968-04-23	2016-06-01	0.0056	ADP.2017-06-14T09:12:09.346

Table A.2 The SN Ia host galaxies observed by MaNGA

SN name	SN RA	SN DEC	IFU RA	IFU DEC	SN time	IFU time	Redshift	IFU ID
SN2018ccl	247.0464	39.8201	247.0482	39.8219	2018-05-28	2015-06-23	0.0268	1-569169
SN2018btb	173.6162	46.3625	173.6187	46.3606	2018-05-14	2016-04-25	0.0338	1-279410
SN2018bbz	261.9680	60.0961	261.9689	60.0973	2018-04-26	2015-09-04	0.0278	1-25680
SN2018ats	153.2345	46.4181	153.2319	46.4177	2018-04-10	2015-03-25	0.0382	1-167380
SN2018aej	236.0959	39.5581	236.0961	39.5590	2018-03-08	2015-06-11	0.0479	1-322806
SN2018ddh	184.6835	44.7820	184.6847	44.7812	2018-07-01	2016-02-16	0.0383	1-258653
SN2017ckx	117.0459	28.2303	117.0457	28.2303	2017-03-28	2015-11-12	0.0272	1-556501
SN2012hj	166.8300	46.3795	166.8320	46.3833	2012-12-04	2015-02-14	0.0246	1-277539
SN2012bm	196.4402	46.4647	196.4440	46.4619	2012-03-27	2015-05-08	0.0248	1-284329
SN2007sw	183.4037	46.4934	183.4036	46.4939	2007-12-29	2015-03-15	0.0257	1-575847
SN2007R	116.6564	44.7895	116.6571	44.7905	2007-01-26	2014-10-29	0.0308	1-339041
SN2006iq	324.8906	10.4849	324.8916	10.4835	2006-09-23	2014-08-31	0.0789	1-114465
SN2006cq	201.1046	30.9563	201.1059	30.9593	2006-05-29	2017-06-15	0.0485	1-575232
SN2003an	201.9731	28.5081	201.9719	28.5082	2003-02-09	2017-03-02	0.0370	1-395622
SN2002aw	249.3711	40.8806	249.3720	40.8799	2002-02-15	2016-03-16	0.0264	1-135668
SN2002G	196.9803	34.0851	196.9784	34.0871	2002-01-18	2017-05-18	0.0336	1-415476
SN2004H	173.4990	49.0629	173.4968	49.0620	2004-01-17	2017-04-16	0.0316	1-576106
PTF11bui	198.2350	47.4535	198.2363	47.4566	2011-04-26	2015-04-15	0.0281	1-285004
PTF11mty	323.5217	10.4235	323.5212	10.4219	2011-09-23	2014-08-31	0.0774	1-114129
PTF12izc	355.8932	0.5687	355.8949	0.5678	2012-09-21	2015-09-16	0.0826	1-29726
PTF13f	247.3591	38.4198	247.3615	38.4194	2013-02-01	2015-06-23	0.0305	1-211264
Gaia15abd	205.2830	23.2830	205.2827	23.2821	2015-02-07	2017-03-01	0.0264	1-568584
SN2017frb	317.9036	11.4974	317.9032	11.4969	2017-07-25	2014-07-04	0.0294	1-113540
SN2019pig	225.3894	49.1095	225.3890	49.1124	2019-09-03	2016-04-28	0.0260	1-246549
SN2019omi	57.2491	0.9269	57.2484	0.9260	2019-08-24	2015-11-06	0.0358	1-229060
SN2017fel	322.3057	−0.2947	322.3060	−0.2948	2017-07-05	2015-09-13	0.0305	1-289846
SN2010dl	323.7540	−0.5133	323.7516	−0.5114	2010-05-24	2016-06-15	0.0302	1-180080
SN2007O	224.0216	45.4047	224.0182	45.4053	2007-01-21	2017-03-04	0.0362	1-576436
SN2006np	46.6645	0.0640	46.6649	0.0620	2006-11-10	2016-11-01	0.1074	1-37863
SN2002ci	243.9081	31.3215	243.9074	31.3213	2002-04-19	2016-05-14	0.0222	1-272321

Appendix B
Tests of the DTD Results

B.1 Algorithm Performance on the Simulated Data

To test the performance of our algorithm on DTD parameter estimate, we designed a test based on the MUSE data used in this paper. Firstly, the MDA model with a given τ and a given s value is used as the fiducial model. Using the fiducial model and the SPAD calculated from `ppxf`, we calculated the fiducial SN rate at every pixel of each galaxy. A coordinate is sampled from each galaxy as the simulated SN explosion site, and the fiducial SN rate is used as the probability of the random sampling.

We used $log_{10}(\tau)$=[7.8,8.0,8.2,8.4,8.6,8.8,9.0,9.2,9.4] and s=[−1.1,−1.2,−1.3, −1.4, −1.5,−1.6,−1.7,−1.8] to create a grid of fiducial DTD models to test our algorithm. The grouping number K is set to be 15, and σ_D is taken from the seeing profile. In Fig. B.1, we show the posterior distribution for τ and s under different fiducial values. In Fig. B.2, we show the estimated τ and s values and errors as a function of the fiducial values.

Castrillo et al. [1] used K-S statistics to recover the DTD parameters from a similar MUSE dataset, which consists of 116 SNe Ia in 102 host galaxies. Moreover, they tested their method from a simulated SN dataset based on MaNGA Pipe3D SPAD data [2]. We notice their pipeline's output is biased to $\tau \sim 100\,\text{Myr}$ and $s \sim -1.1$ for most of the fiducial $\tau \in [50, 1000]\,\text{Myr}$, $s \in [-0.6, -1.8]$, and the bias is observed in different sample sizes from $N = 20$ to $N = 500$. From our simulation results, the recovered $log_{10}(\tau)$ value is larger than the fiducial $log_{10}(\tau)$ by $\sim 0.2\,\text{dex}$ when the fiducial value is smaller than 8.6, and the recovered s value is smaller than the fiducial s value by 0.3 for some of the fiducial s bins. To conclude, our DTD estimation method based on K-means and EMD grouping algorithm did not show significant bias in the simulation results. In contrast, the simulations in [1] show larger bias effects, and we surmise this phenomenon can be attributed to two reasons: (1) The simulation in [1] is based on MaNGA, and the data signal-to-noise ratio is lower than that from MUSE. (2) The DTD estimation method in [1]

X. Chen, *Deep Learning on Type Ia Supernovae*, Springer Theses,
https://doi.org/10.1007/978-3-032-13032-7

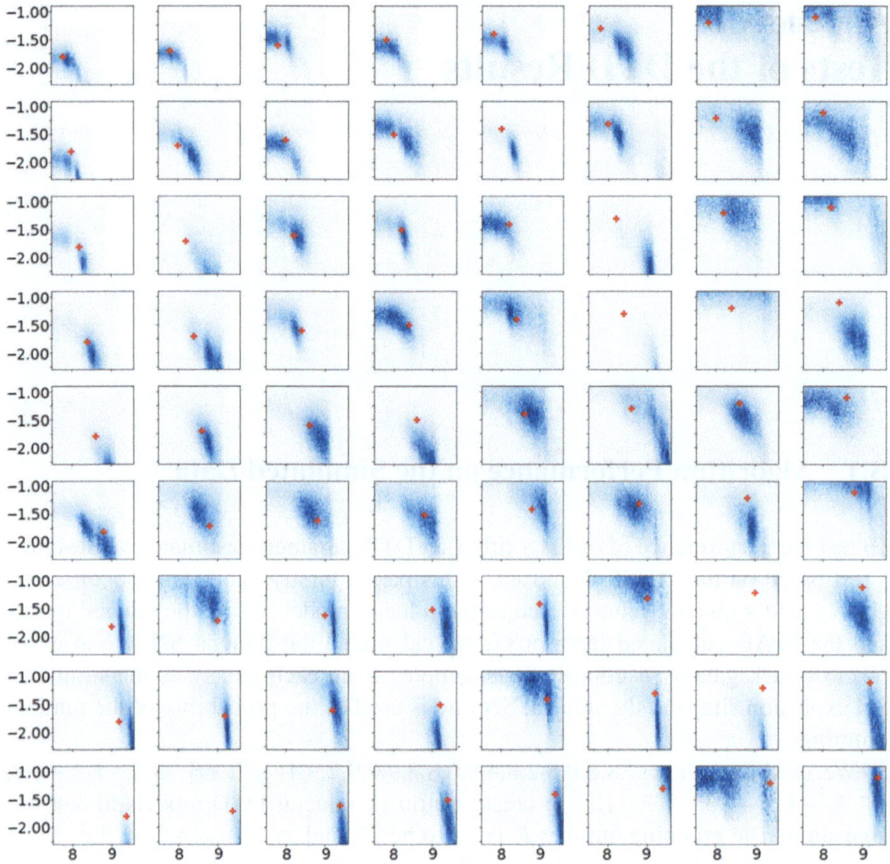

Fig. B.1 The DTD parameter estimates from the simulated data. Red crosses are the fiducial values used to generate the simulated data. Blue shadowed regions are the posterior distribution from the algorithm. X-axis is $\log_{10}(\tau)$, and y-axis is s

is based on K-S test, which is a nonparametric test with different assumptions from our maximum likelihood estimate method.

B.2 Results from Pipe3D Data

To further test our algorithm's performance and the reliability of `ppxf` SPAD, we use the SPAD data calculated by Pipe3D for the MaNGA data, to estimate the DTD parameters. The `Pipe3D` pipeline is introduced in [3]. The implementation of `Pipe3D` onto the MaNGA data is discussed in [2].

In Fig. B.3, we compared the mass-averaged stellar ages derived from `ppxf` and that from `Pipe3D` for six example galaxies, and the 21×21 spatial pixels at the

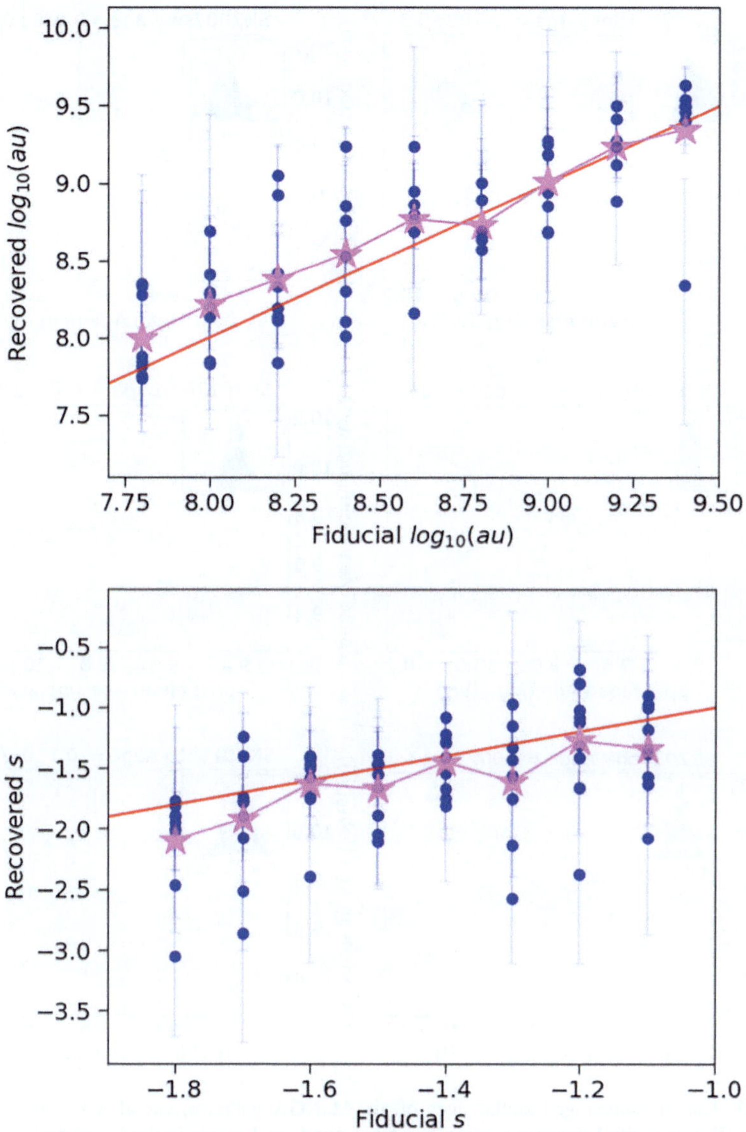

Fig. B.2 Left panel: the fiducial τ values used in simulation and the recovered τ values from the maximum likelihood estimate with the galaxies grouped into $K = 15$. Right panel: the fiducial s values and the recovered s values. The data used in the figures are from Fig. B.1. The blue transparent error bars are the 1σ limits. The violet stars are the means of the recovered parameter values. The red line shows the ideal case that the fiducial value being equal to the recovered value, as a reference

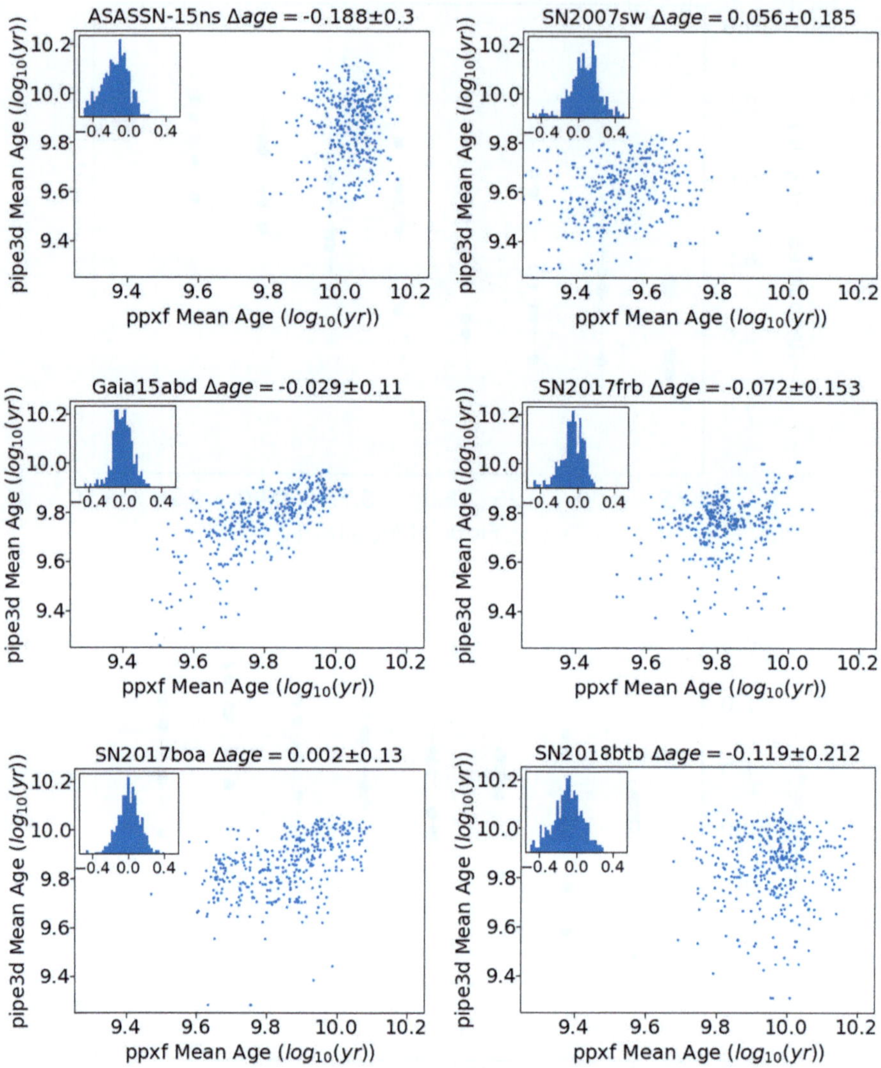

Fig. B.3 The mass-averaged stellar ages of six MaNGA galaxies, calculated from `ppxf` and `Pipe3D`. X-axis is the logarithmic mean age from `ppxf`, and y-axis is the logarithmic mean age from `Pipe3D`. The inset of each figure is the histogram of the difference between the two mass-averaged stellar ages. The supernova names in the galaxies are labeled in the titles. The mean age deviation is shown in the title

center of each data cube are included. We notice the correlation between the mass-averaged stellar ages is weak, this could be due to the low signal-to-noise ratio in most of the pixels in the MaNGA data cube which introduce extra uncertainties in calculating the SPADs, and most of the selected galaxies show a similar age between 10^9 and $10^{1}0$ years. Sánchez et al. [3] compared the luminosity-averaged stellar ages

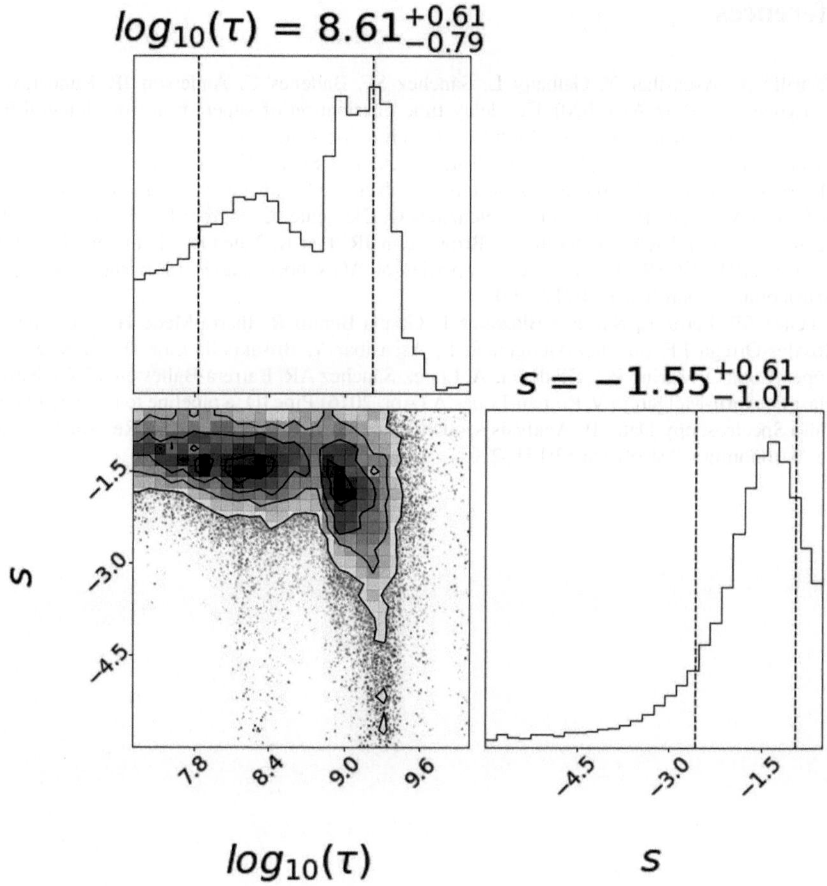

Fig. B.4 The posterior probability distribution of the MDA model parameters using $K = 15$ groups for the MaNGA data and Pipe3D SPAD

from Pipe3D, Starlight, Steckmap, and absorption line indices. They found the age bias is between −0.06 and 0.06 dex, and the age deviation is between 0.15 and 0.22 dex. In our comparison results, most of the MaNGA galaxies show similar age deviation and bias, and thus we conclude the SPADs from ppxf are consistent with that from Pipe3D.

In Fig. B.4, we used the SPAD from Pipe3D to estimate the DTD parameters in MDA model. The grouping number is $K = 15$, and σ_D is taken from the seeing profile. We notice the result is in agreement with that using the SPAD from ppxf, which is shown in Fig. 5.7.

References

1. Castrillo A, Ascasibar Y, Galbany L, Sánchez SF, Badenes C, Anderson JP, Kuncarayakti H, Lyman JD, Díaz AI (2020) The delay time distribution of supernovae from integral-field spectroscopy of nearby galaxies. Mon Not R Astron Soc staa3876
2. Sánchez SF, Avila-Reese V, Hernandez-Toledo H, Cortes-Suárez E, Rodríguez-Puebla A, Ibarra-Medel H, Cano-Díaz M, Barrera-Ballesteros JK, Negrete CA, Calette AR, de Lorenzo-Cáceres A, Ortega-Minakata RA, Aquino E, Valenzuela O, Clemente JC, Storchi-Bergmann T, Riffel R, Schimoia J, Riffel RA, Rembold SB, Brownstein JR, Pan K, Yates R, Mallmann N, Bitsakis T (Apr. 2018) SDSS IV MaNGA - properties of AGN host galaxies. Revista Mexicana de Astronomía y Astrofísica 54:217–260
3. Sánchez SF, Pérez E, Sánchez-Blázquez P, García-Benito R, Ibarra-Mede HJ, González JJ, Rosales-Ortega FF, Sánchez-Menguiano L, Ascasibar Y, Bitsakis T, Law D, Cano-Díaz M, López-Cobá C, Marino RA, Gil de Paz A, López-Sánchez AR, Barrera-Ballesteros J, Galbany L, Mast D, Abril-Melgarejo V, Roman-Lopes A (Apr. 2016) Pipe3D, a pipeline to analyze Integral Field Spectroscopy Data: II. Analysis sequence and CALIFA dataproducts. Revista Mexicana de Astronomía y Astrofísica 52:171–220

MIX
Papier aus verantwortungsvollen Quellen
Paper from responsible sources
FSC® C105338

If you have any concerns about our products,
you can contact us on
ProductSafety@springernature.com

In case Publisher is established outside the EU,
the EU authorized representative is:
Springer Nature Customer Service Center GmbH
Europaplatz 3, 69115 Heidelberg, Germany

Printed by Libri Plureos GmbH
in Hamburg, Germany